Arc Welding

Eighth Edition

by

John R. Walker • W. Richard Polanin

Publisher

The Goodheart-Willcox Company, Inc.

Tinley Park, Illinois

www.g-w.com

Cover Image: © R. Maisonneuve/Photo Researchers, Inc.

Library of Congress Cataloging-in-Publication Data

Walker, John R.
    Arc welding / by John R. Walker, and W. Richard Polanin.
        p. cm.
    ISBN 978-1-60525-189-9
    1. Electric welding. I. Polanin, W. Richard. II. Title.

TK4660.W27 2010
671.5'212—dc22   2009020995

Using This Write-in Text

Arc Welding presents the fundamentals of this skilled trade in an easy-to-understand manner. Each unit includes problems for you, the student welder, to solve.

Arc welding requires a great deal of practice to produce satisfactory weldments. To become a good welder, you will have to spend many hours running beads and making joints presented in this write-in text. Read the text matter carefully, study and complete the questions, and perform the activities as required. This will aid you in developing skills and techniques that will enable you to enter the welding industry and to prepare yourself for many exciting career opportunities.

Each practice piece should be evaluated and every effort should be made to correct any problem encountered. Where possible, test each weld for integrity, penetration, and appearance. Your instructor will help you to do this. To conserve material, it is recommended that both sides of the practice piece be used whenever practical. Identify your work by stamping your name or initials on each practice piece.

The questions in Check Your Progress at the end of each unit will help you determine how well you understand the related information that the welder must know. Additional up-to-date welding knowledge can be acquired by carrying out the suggested activities. Be sure to practice safety at all times.

About the Authors

John R. Walker is the author of thirteen textbooks and has written many magazine articles. Mr. Walker did his undergraduate studies at Millersville University and has a master of science degree in industrial education from the University of Maryland. He taught industrial arts and vocational education for thirty-two years and served as the Supervisor of Industrial Education for five years. He also worked as a machinist for the U.S. Air Force and as a draftsman at the U.S. Army Aberdeen Proving Grounds.

W. Richard Polanin is a professor at Illinois Central College, as well as the coordinator of the Manufacturing Engineering Technology and Welding Technology programs. Dr. Polanin has a bachelor's degree and a master's degree from Illinois State University, as well as a doctorate degree from the University of Illinois. In addition to his twenty-five years of teaching experience, he is an active consultant in welding and manufacturing. He is an AWS-certified welding inspector, an AWS-certified welding educator, and an SME-certified manufacturing engineer. He has published numerous technical papers and has made many technical presentations in the areas of welding, manufacturing, robotics, and manufacturing education.

Contents

Unit 1

Introduction to Arc Welding

After completing this unit, you will be able to:
○ Explain how the arc is produced for welding purposes.
○ List the three most popular commercially used arc welding methods.
○ Define each of the three popular metal joining processes.

Dozens of different welding processes are used by modern industry, **Figure 1-1.** However, arc welding, gas welding, and resistance welding are by far the most common. This write-in textbook is concerned with arc welding and the basic processes involved.

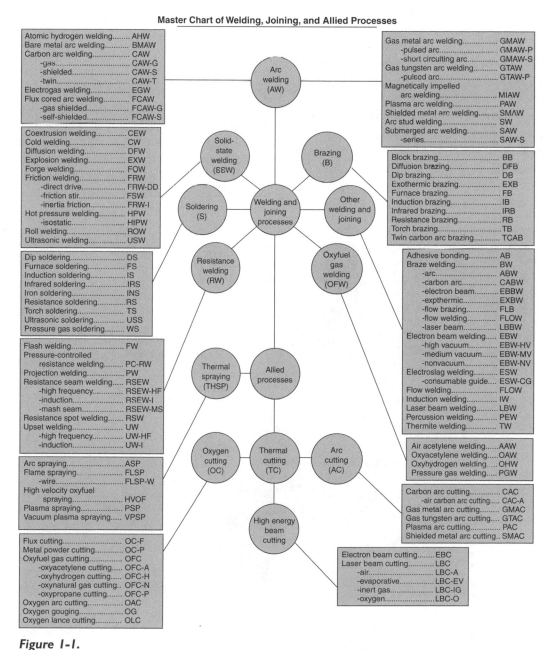

Figure 1-1.
A chart of the many welding processes used in industry. (Adapted from AWS A3.0:2001, Figures 54A and B, Master Chart of Welding and Joining Processes and Master Chart of Allied Processes, reproduced with permission from the American Welding Society, Miami, FL)

A weld is made when two or more metal pieces are melted along a common edge or surface. The pieces are fused together when the molten metal cools and solidifies.

In *arc welding,* **Figure 1-2,** electric current is used to generate an arc between an electrode and the work. A very high temperature (about 8000°F) is created at the point of the arc. For most arc welding processes, the tip of the electrode and a small portion of the work become molten in the intense heat. When the arc is moved or stopped, the molten metal fuses into a single piece. It is difficult to name a manufacturing industry that does not depend on arc welding in some portion of its work.

There are several arc welding methods. The most popular commercially used methods are *gas metal arc welding (GMAW), gas tungsten arc welding (GTAW),* and *shielded metal arc welding (SMAW).* GMAW uses a continuous wire feed from a coil of wire as the electrode that produces the arc. In GTAW, a non-melting electrode made of tungsten is used to produce the arc. Because SMAW uses a covered wire electrode as shown in Figure 1-2, the process is also called *stick welding.*

Welding, Brazing, and Soldering

Many workers who are not familiar with various welding processes think that welding, brazing, and soldering are all the same. They are not. You, as a student welder, must be familiar with these metal joining processes and know how they differ.

Welding is the process in which metal pieces are joined by heating them to a temperature high enough to cause them to melt and fuse together into a single piece. This may be accomplished with or without the application of pressure, and with or without the use of filler metal similar in composition and melting point to the base metals being joined. Welding always takes place at a temperature above 800°F (412°C).

Brazing is a process in which metal pieces are joined by heating them to a suitable temperature above 840°F (450°C) but below their melting points, and adding nonferrous filler metal having a melting point below that of the base metals.

Soldering is a process that uses alloys with low melting points (under 840°F [450°C]) to join the metal parts. The molten solder adheres to the metal surfaces and fills the space between the parts to be joined. A soldered joint is not as strong as a welded or brazed joint. Not all metals can be soldered.

Figure 1-2.
Arc welding is a highly skilled occupation. In addition to having the ability to make good welds, the welder must also have considerable knowledge of the characteristics of metals. (Miller Electric Mfg. Co.)

Name _____   Score _____

Check Your Progress

1. Welding, brazing, and soldering are not the same. In your own words, describe each process and explain how they differ.

   Welding: _____

   _____

   _____

   _____

   _____

   _____

   _____

   Brazing: _____

   _____

   _____

   _____

   _____

   _____

   Soldering: _____

   _____

   _____

   _____

   _____

   _____

2. In the following chart, list objects you use that were fabricated (assembled) by welding or contain parts fabricated by welding.

Object	Parts Fabricated by Welding

3. Obtain objects fabricated by welding and bring them to class.

4. Explain how the arc is produced for GMAW, GTAW, and SMAW.

Notes

Unit 2
Measurement in Welding

After completing this unit, you will be able to:

○ Identify measuring tools commonly used by the welder.

○ Accurately label the measurements on a rule with 1/16″ graduations.

○ Produce exact measurements using a U.S. Conventional rule.

○ Produce exact measurements using a metric rule.

It is very important that the pieces to be welded are cut to correct size and shape. Many times the welder must do the cutting. To make the cut quickly and accurately, you must be able to read a rule.

Shown in **Figure 2-1** are several of the measuring tools the welder is expected to be able to use.

Most measurement in the United States is still made in the U.S. Conventional system, but use of SI metric (the standard for the rest of the world) is growing. For general welding work, you should be able to read a U.S. Conventional rule to 1/16″ (one-sixteenth inch), or a metric rule to 1 mm (one millimeter).

Secure a rule or tape measure and examine the edges used to measure. Each mark or division on the edge of the rule is called a **graduation.** Look carefully at the edge used to measure to 1/16″. Many rules have 16 or 1/16 stamped or engraved on this edge. If your rule does not, count the graduations. There will be 16 of them per inch, **Figure 2-2.**

Figure 2-2.
Section of rule with 1/16″ graduations.

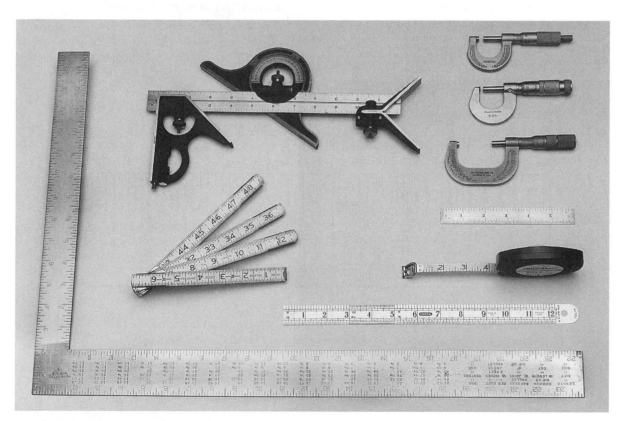

Figure 2-1.
Measuring tools with which the welder should be familiar. How many can you name correctly?

Note that the 1″ graduation is the longest, the 1/2″ graduation is next longest, and so on down to the 1/16″ graduation, which is the shortest.

Until you are more familiar with the rule, visualize or imagine that each 1/16″ division is numbered as shown in **Figure 2-3.**

You may also find it easier to count the graduations or divisions when you first start to use the rule. However, after some practice this should not be necessary.

One thing to remember, fractional measurements are always reduced to the lowest terms. A measurement of 8/16″ would be read as 1/2″; 2/16″ as 1/8″; etc. However, most shop drawings use decimal-inch measurements. Therefore, you should be able to convert fractions to decimals. There is a conversion table in the reference section of this textbook. Try to become familiar with the conversions of the common measurement fractions (1/8″ = 0.125″, 1/4″ = 0.25″, etc.).

Now, secure a rule with metric graduations. It will usually be marked with the letters "cm" (centimeters) or "mm" (millimeters). Commonly, as shown in **Figure 2-4,** the numbered divisions will be centimeters. Each centimeter is divided into 10 millimeters. The millimeter is the most common metric unit used in industry and is the smallest unit you will use in welding work.

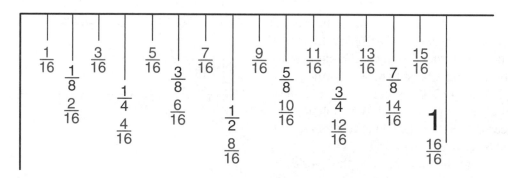

Figure 2-3.
Until you are familiar with measuring to 1/16″, visualize each inch of the rule numbered as shown above.

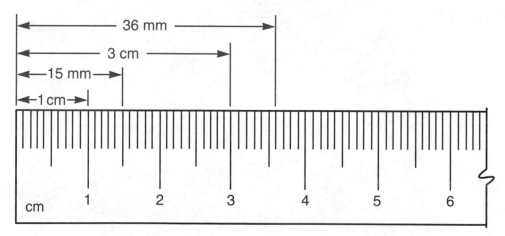

Figure 2-4.
A typical metric rule with millimeter (mm) and centimeter (cm) divisions. Metric measurements are always stated as whole numbers. Therefore, you would say "15 millimeters" instead of "one and one-half centimeters."

Name _____ Score _____

Check Your Progress

Problem 1: How many can you answer correctly? Use the appropriate space for the answer.

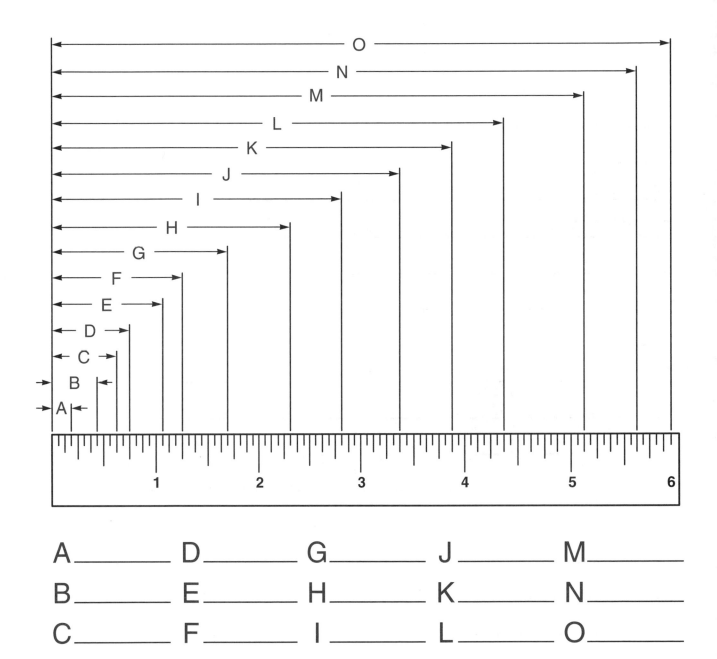

A_____ D_____ G_____ J_____ M_____

B_____ E_____ H_____ K_____ N_____

C_____ F_____ I_____ L_____ O_____

Problem 2: Secure two rules: one with 1/16″ graduations and one with 1 mm. Use the U.S. Conventional rule to measure lines A–J, and the metric rule to measure lines K–T. Place your answer in the blank to the left of each line, as shown in the example. Reduce fractions to their lowest common denominator (for example, 6/8 = 3/4). Also, convert the fractions to two-place decimals using the conversion chart on page 182 of this text (for example, 1/2 = .50).

Example $\dfrac{7}{8}$ ⊢————⊣

A ____ ⊢——————————⊣

B ____ ⊢————⊣

C ____ ⊢—————————⊣

D ____ ⊢————————————⊣

E ____ ⊢——————————————————⊣

F ____ ⊢—————————————⊣

G ____ ⊢———————————————————⊣

H ____ ⊢————————————————————————————⊣

I ____ ⊢———————————————————⊣

J ____ ⊢———————————————————————⊣

K ____ ⊢————————————————————————————⊣

L ____ ⊢—————————⊣

M ____ ⊢———————————————————⊣

N ____ ⊢—————————————————————————⊣

O ____ ⊢——————————————————————————⊣

P ____ ⊢————————————————————————⊣

Q ____ ⊢——————————————————————————————⊣

R ____ ⊢————————————————————————⊣

S ____ ⊢————————⊣

T ____ ⊢————————————————⊣

Name _____ Score _____

Problem 3: Use a U.S. Conventional rule to measure lines A–N, and a metric rule to measure lines O–Y. Place your answer in the appropriate blank. Reduce fractions to their lowest common denominator.

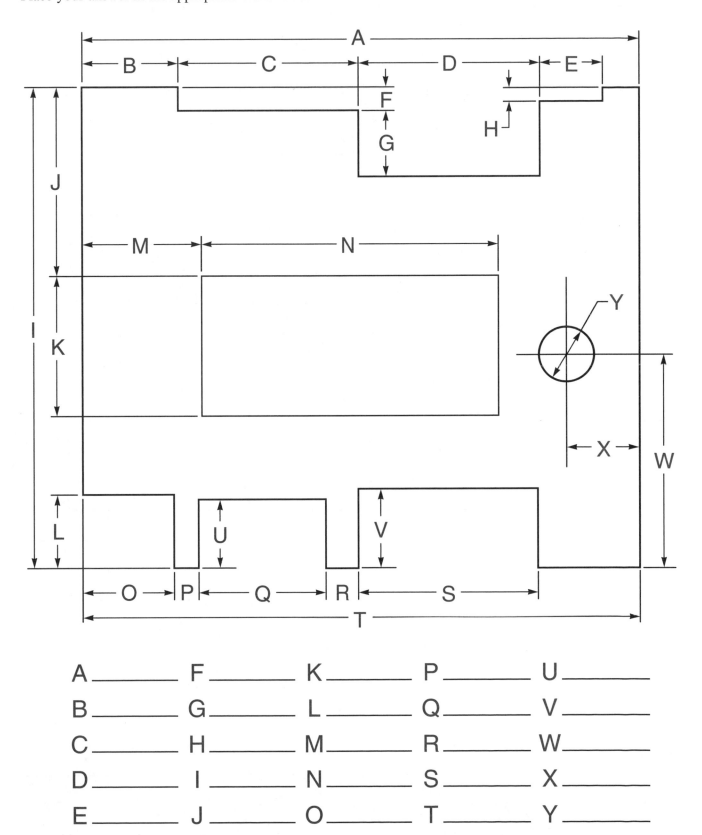

A_____ F_____ K_____ P_____ U_____

B_____ G_____ L_____ Q_____ V_____

C_____ H_____ M_____ R_____ W_____

D_____ I_____ N_____ S_____ X_____

E_____ J_____ O_____ T_____ Y_____

Notes

Unit 3
Welding Safety

After completing this unit, you will be able to:

○ List the safety equipment a welder should use regularly.

○ Explain the precautionary purpose for each of the safety items presented.

○ Identify safety precautions for the environment around the welding area.

The welder must be aware of the hazards of the trade. The welder is involved with many potentially dangerous activities during the working day.

- Hot metal is handled.
- High current electricity is used.

Figure 3-1.
The arc welder can protect against many welding hazards by dressing properly for the job. Note the protective devices this welder is wearing. (Miller Electric Mfg. Co.)

- Toxic fumes are generated during the welding operation.
- Large quantities of ultraviolet rays are emitted by the arc that can cause severe eye damage.
- Droplets of molten metal often fly from the arc.
- Heavy pieces of metal must sometimes be positioned for welding.
- High noise levels from welding equipment and other workplace equipment.

To remain injury free, the welder must take a positive approach to safety by observing basic safety precautions.

A Positive Approach to Safety

Welders can protect themselves from many arc welding hazards by dressing properly for the job, **Figure 3-1.** A *head shield,* also called a *welding helmet* or a *welding hood,* will protect the face and eyes from the rays of the electric arc and the spatter of molten metal. See **Figure 3-2.**

Figure 3-2.
Many types and styles of welding helmets are available. The latest designs have an auto-darkening unit that reacts in 1/25,000 of a second and continuously adjusts from shades 9 to 12. The lens is clear until the arc is struck. (UVEX Safety, Inc.)

Shields are fitted with special darkened lenses to reduce the ultraviolet and infrared rays that reach the welder's eyes. A covering of inexpensive clear glass protects the dark filter glass from the small particles of molten metal that fly about during the welding operation, **Figure 3-3.** The darkness of the lens is indicated by a number and should be selected based on the amount of amperage used for welding. For the practice welds in this text, a number 10 lens will provide appropriate protection.

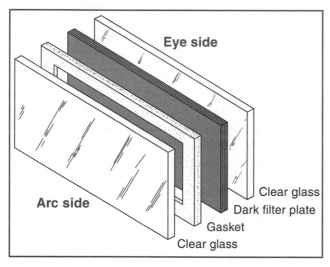

Figure 3-3.
This illustration shows how the lenses are fitted in the welding helmet. The clear cover glass is inexpensive and should be replaced if it becomes cracked or covered with weld spatter.

Never use a shield that has a cracked lens. Also, wear safety goggles under the shield. Never arc weld or watch arc welding being done without using a protective shield. Gas welding goggles or sunglasses will not protect your eyes from the ultraviolet and infrared rays emitted by the arc.

Most welders prefer the head shield to the hand-held shield because the head shield permits both hands to be free for welding.

Gauntlet-type gloves, a leather jacket or a leather apron and sleeves will provide protection from the molten metal sparks and spatter. Wear high-top leather shoes (with safety tips) rather than the low oxford-type shoes. Never wear canvas shoes while welding.

If leather protective clothing is not available, clothing made from heavy, fire resistant cloth may be worn. The clothing should be tight fitting. Trousers should *not* have cuffs, since the cuffs may catch burning particles as they fall. Never carry easily ignited materials, such as matches, lighters, plastic combs, or pens, in your pocket.

Check Your Equipment

Carefully examine your equipment before starting. Frayed cables with loose lugs, damaged electrode holders, or damaged ground clamps should be repaired or replaced.

Hot Metal

Use tongs or pliers to handle hot metal. Cool hot metal in a quench tank or place it out of the way to cool so that others will not be burned.

Treatment of Burns

Welded metal remains hot for an extended length of time. Therefore, you should consider all material in your work area hot. If you do pick up a hot piece of metal or if some spatter finds its way through your protective clothing, you may have to treat the burn.

Most minor burns require only simple first aid and often heal after a few days. However, if you think the burn is more severe, you should consult with a health care professional. Many schools have a nurse on staff who can provide immediate treatment or provide a referral.

The first step in treating a minor burn is to remove any clothing from the site of the burn. Then use a cool cloth on the burn area. To relieve pain, apply soothing lotions designed for burn treatment. The lotions can often be found at your local pharmacy. Finally, to help prevent infection, cover the burn site with a dry, clean cloth.

Toxic Fumes

The fumes developed by the electrode covering and hot metal may contain poisonous metal oxides. These fumes can be avoided by welding in a well-ventilated area or by using forced ventilation. If the area cannot be ventilated, *do not weld*.

Welding Area

Shield the welding operation if others are working in the area. This will reduce the possibility of a fellow worker accidentally looking at the arc and injuring his or her eyes.

The flying sparks that accompany arc welding are also a serious hazard. In addition to burning the welder, they can start fires. Keep the welding area free of flammable materials and solvents.

Never weld a tank or container of any type until you determine whether it has contained flammable liquids. If it did, get the tank steam cleaned or fill it with water before welding. Refrain from arc welding in damp or wet areas.

Get help to position heavy metal plates for welding. Do not attempt to move them by yourself.

Never enter a confined or enclosed space without proper authorization. Welding will use up oxygen rapidly and produce fumes that displace air. Without proper precautions, you could suffocate.

Noise

Many welding areas are quite noisy. If not protected, your hearing could eventually be impaired. Invest in and wear an approved-type hearing protector, **Figure 3-4.**

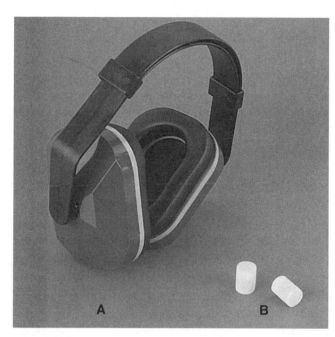

Figure 3-4.
To protect your hearing, wear an approved-type hearing protector. A—Headphone-type hearing protector. B—Earplugs.

Safety Reminders

You must be familiar with situations in a welding facility that can endanger your safety, and the safety of others. The best way to eliminate or control the safety hazards in a welding shop is to eliminate the conditions that are potentially dangerous. Here are a few general safety warnings that pertain to welding:

- Ultraviolet rays are harmful to eyes. Safety goggles or a helmet with proper lenses will provide protection, **Figure 3-5**.
- Ultraviolet radiation markedly increases the possibility of eye cataracts. Even small amounts of radiation may cause cataracts. Welders, as well as helpers and other people in the area, must take precautions at *all* times to prevent eye injury by wearing approved welding eye protection.
- Always protect your eyes, skin, and respiratory system. Provide adequate ventilation during all welding operations.
- Coatings on metals can be a problem when the metal is heated. Do not breathe the fumes generated during the welding process.
- Red lead is sometimes used to protect metals exposed to the elements. Lead oxide fumes produced by welding or burning operations can produce lead poisoning.
- Cadmium plating is commonly used on small parts. Cadmium fumes, even at low levels, can be harmful to your lungs and liver.
- Terne plating is sometimes used on steel. Terne is a metallic lead coating that is dangerous when heated.
- Fluorides are common in fluxes. Their fumes can be dangerous if ventilation is poor.

Shielded Metal Arc Welding Shade Lens			
Electrode Diameter	Current (Amps)	Minimum Shade Number	Recommended Shade Number
<3/32" (2.4 mm)	<60	7	8
3/32"–5/32" (2.4 mm–4.0 mm)	60–160	8	10
5/32"–1/4" (2.4 mm–4.0 mm)	160–550	10	12
>1/4" (4.0 mm)	250–550	11	14

Figure 3-5.
Always start with a lens shade darker than the suggested shade number. After a test weld, adjust the lens shade to a lighter shade that gives an adequate view of the weld area. Do *not* use a lens shade below the minimum protective shade number.

- Beryllium is toxic in even small amounts. Any operation involving beryllium must be contained. Follow approved methods when working with this metal.
- Oil smoke is often a problem. The fumes produced can be dangerous. Be sure the work area is well ventilated.
- When in doubt about the proper and safe procedures to follow in doing a job, contact trained safety personnel. Never attempt a job until you are sure you will not injure yourself or other people in the area.
- Zinc is used to prevent steel from rusting. The coating process is called galvanizing. Breathing the zinc can be harmful to your lungs and cause you to have flu-like symptoms.

Name _____ Score _____

Check Your Progress

1. Welders are exposed to many hazards in their jobs. List five of them.

   a. _____

   b. _____

   c. _____

   d. _____

   e. _____

2. The head shield protects the welder from:

   Check the correct answer(s).

   a. _____ rays from the welding arc.

   b. _____ flying molten metal.

   c. _____ Both of the above.

   d. _____ None of the above.

3. Why should the welder's pants *not* have cuffs?

   _____

   _____

4. What type of shoes should the welder wear?

   _____

5. What type of shoes should a welder *not* wear?

   _____

6. How can toxic fumes be kept to a minimum when welding? Check correct answer(s).

   a. _____ Use special welding rods.

   b. _____ Weld only in well-ventilated areas.

   c. _____ Use forced ventilation if needed.

   d. _____ All of the above.

   e. _____ None of the above.

7. The welder should use _____ to _____ handle hot metals.

8. The welding area should be kept clear of _____ materials because _____ _____ can start fires.

9. The welding area should be shielded to reduce the possibility of a fellow worker _____.

10. Get help to position _____ _____ _____ for welding.

Things to Do

1. Examine all the head shields in the shop. Replace lenses that are cracked or covered with weld spatter. Clean the headbands and repair those that are damaged.

2. Examine and repair or replace other welding equipment that is torn, cut, or frayed. Dispose of leather safety equipment that is soaked with oil.

3. Clean welding areas of solvents, oils, and flammable articles. Check ventilating equipment. Lubricate fan motors.

4. Make storage facilities for equipment that cannot be properly stored in present facilities.

5. Research the different classes of fires and fire extinguishers. Report your findings to the class.

Unit 4
Common Types of Welds and Joints

After completing this unit, you will be able to:
○ Identify the four different weld types.
○ Name the five types of basic weld joints.
○ List six of the groove styles presented.

Welds

Each welding job you do will require one or more of the four welds shown in **Figure 4-1.**

- The **bead weld** is composed of a narrow layer or layers of metal deposited in an unbroken string on the surface of the metal. The bead weld is often called a ***stringer bead.***
- The ***groove weld*** is a weld made between the two pieces of metal to be joined.
- The ***fillet weld*** is approximately triangular in shape and is used when joining two surfaces at an angle.
- The ***plug weld*** is a weld made *through* one piece of metal to join it to another metal surface.

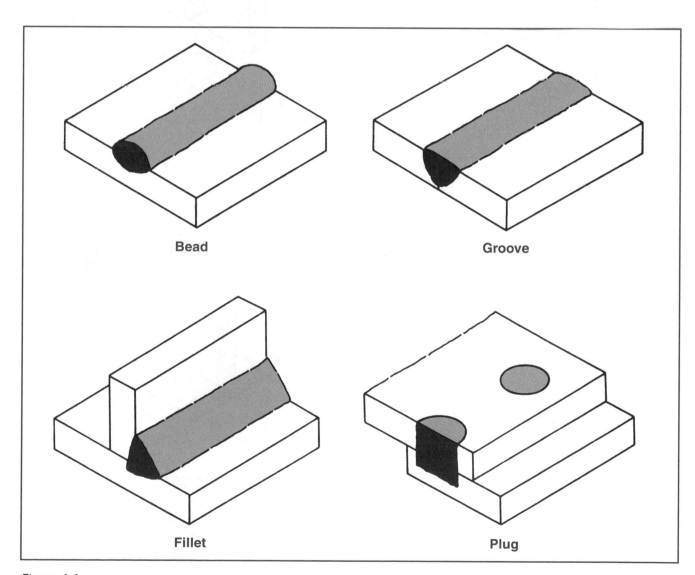

Bead

Groove

Fillet

Plug

Figure 4-1.
Each welding job will make use of one or more of these weld types.

Joints

There are five basic joints (ways of arranging the metal pieces in relation to each other so they can be welded) used in arc welding, **Figure 4-2.** Other joints used in arc welding are combinations and variations of these basic joint designs.

Grooves

To ensure a solid weld, it is frequently necessary to combine the joint with one of the groove styles shown in **Figure 4-3.** The groove is the opening provided between the two metal pieces to be joined by a groove weld.

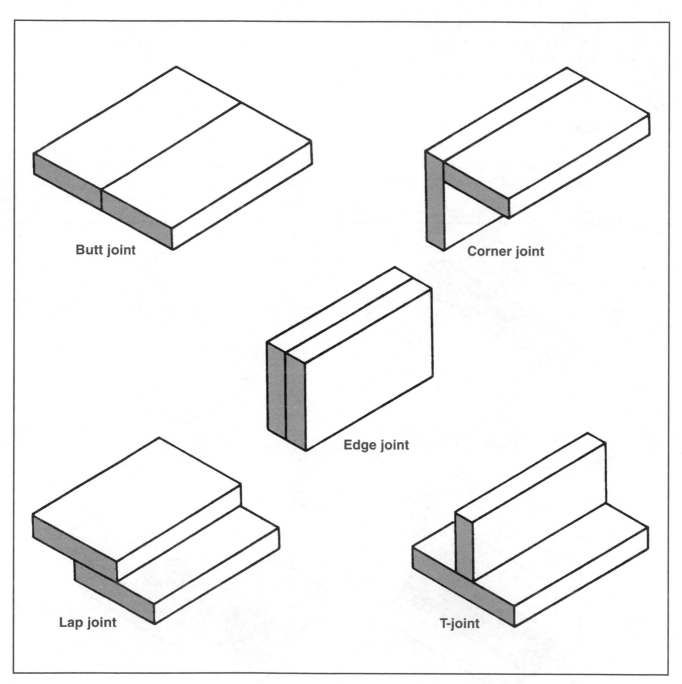

Figure 4-2.
Basic weld joints.

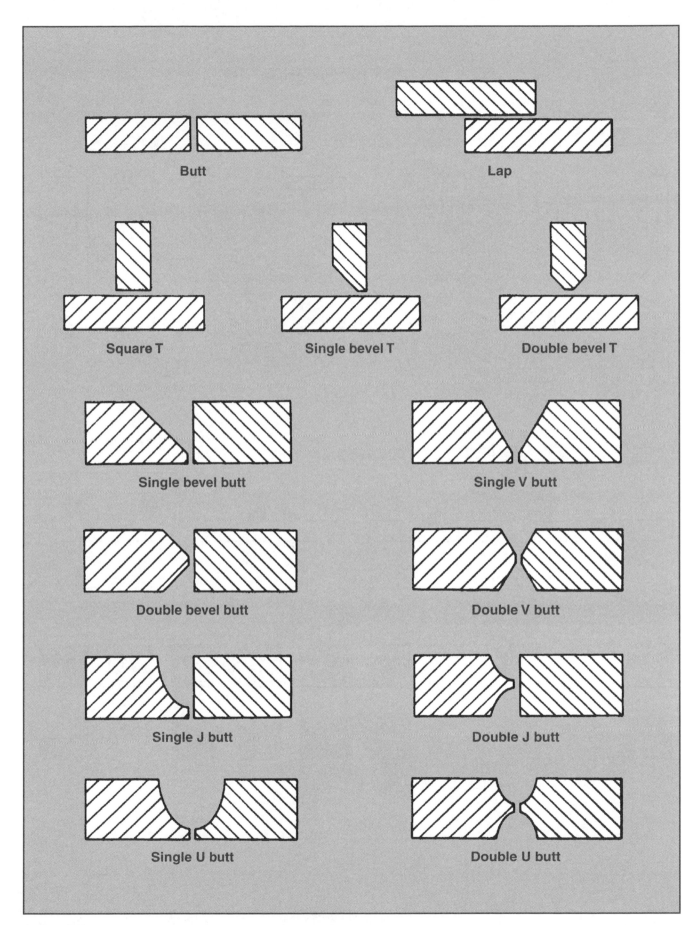

Figure 4-3.
Grooves are frequently required to ensure solid welds.

Name _____ Score _____

Check Your Progress

Problem: Identify the following types of welds and joints.

Joints

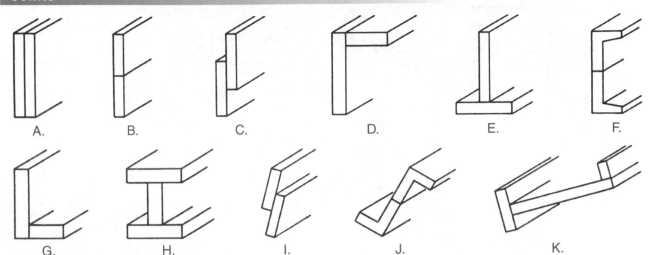

A.   B.   C.   D.   E.   F.

G.   H.   I.   J.   K.

Welds

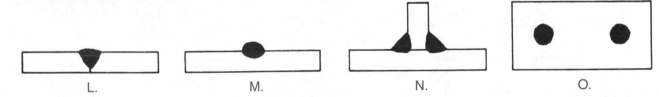

L.   M.   N.   O.

Joint Preparation

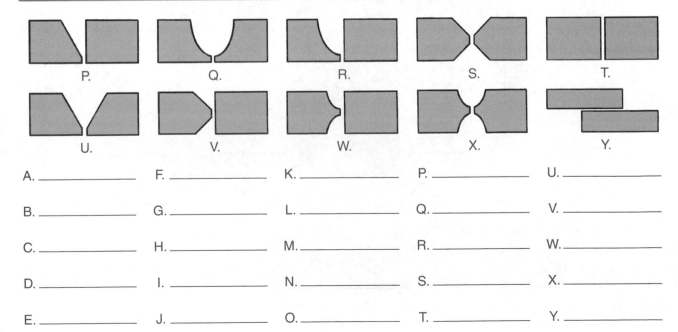

P.   Q.   R.   S.   T.

U.   V.   W.   X.   Y.

A. _____   F. _____   K. _____   P. _____   U. _____

B. _____   G. _____   L. _____   Q. _____   V. _____

C. _____   H. _____   M. _____   R. _____   W. _____

D. _____   I. _____   N. _____   S. _____   X. _____

E. _____   J. _____   O. _____   T. _____   Y. _____

Things to Do

1. Make a collection of small metal parts that have been fastened together by arc welding. Bring these to class and discuss the types of welds used. Try to determine why a particular type weld was used for each situation.

2. Using catalogs, magazines, and other source materials, clip out pictures of products that would require the use of welds presented in this unit. Prepare a bulletin board display indicating how the welds were used.

3. Visit a local welder or welding contractor and ask about the many ways common types of welds are used in the daily routines.

Notes

Unit 5
Welding Symbols

After completing this unit, you will be able to:
- ⭕ Identify the weld type indicated by the standard symbols presented.
- ⭕ Determine the correct weld size and side of the joint based on the welding symbol.
- ⭕ Sketch welding symbols representing welds on an actual part.

Some method had to be devised to tell the welder what type of weld the engineer wanted on the job. The art of welding reached the point where more information was needed than "heavy weld all joints." The *American Welding Society (AWS)* developed and standardized the basic welding symbols shown in **Figure 5-1.** These symbols provide a means for giving complete and specific welding information on the drawing of the part to be welded.

The welding symbol placed on the drawing is made up of basic and supplementary weld symbols that have been selected to describe the required weld, **Figure 5-2.** Typical welding symbols are shown in **Figure 5-3.**

Groove							
Square	Scarf	V	Bevel	U	J	Flare-V	Flare-bevel

Fillet	Plug	Slot	Stud	Spot or projection	Seam	Back or backing	Surfacing	Edge

Figure 5-1.
Weld symbols. These are part of the complete welding symbol. (AWS A2.4:2007, Figure 1, Weld Symbols, reproduced with permission from the American Welding Society, Miami, FL)

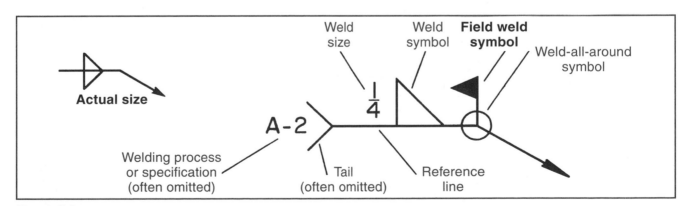

Figure 5-2.
The welding symbol is made up of basic and supplementary weld symbols and gives complete and specific welding information to the welder.

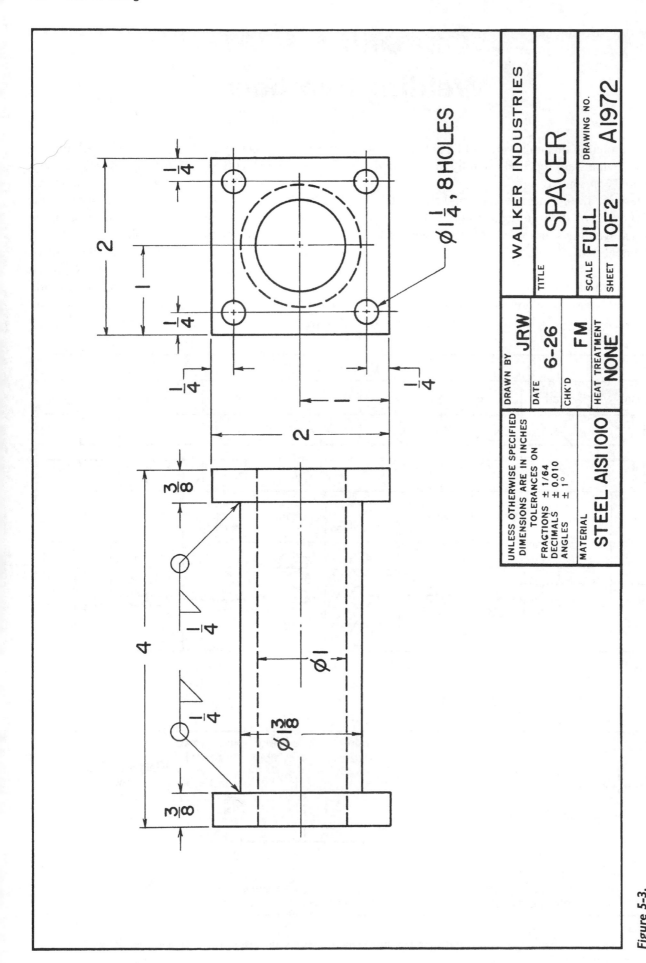

Figure 5-3.
This drawing showing the use of welding symbols.

How Welding Symbols Are Used

When using welding symbols to specify the required weld, the location of the *arrow* in relation to the joint is very important. When the weld is to be made on the *arrow side* of the joint, the weld symbol is placed on the reference line as shown in **Figure 5-4.**

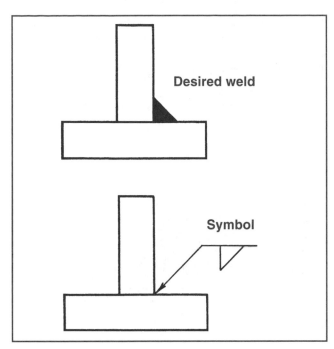

Figure 5-4.
This weld symbol indicates the weld is to be made on the *arrow side* of the joint. Note the weld symbol is *toward* the reader.

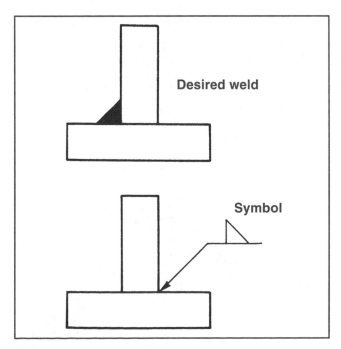

Figure 5-5.
This weld symbol indicates that the weld is to be made on the *other side* of the joint.

The symbol is *toward the reader.* A weld on the other side of the joint is specified by placing the weld symbol on the reference line *away from the reader,* **Figure 5-5.**

Weld size is indicated as shown in **Figure 5-6.** The *field weld* symbol, **Figure 5-7,** means the weld is to be made when the parts are assembled away from the shop. Typical welding symbols and their meanings are shown in **Figures 5-8** and **5-9.** Study them carefully.

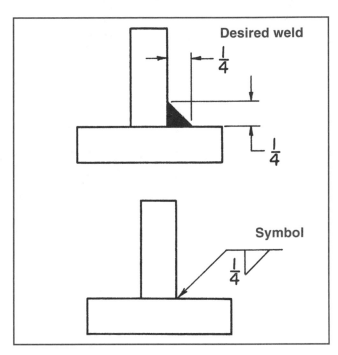

Figure 5-6.
Symbol indicating size of weld.

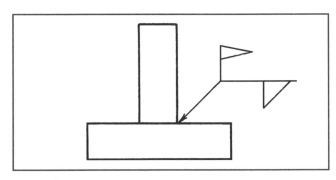

Figure 5-7.
Field weld symbol.

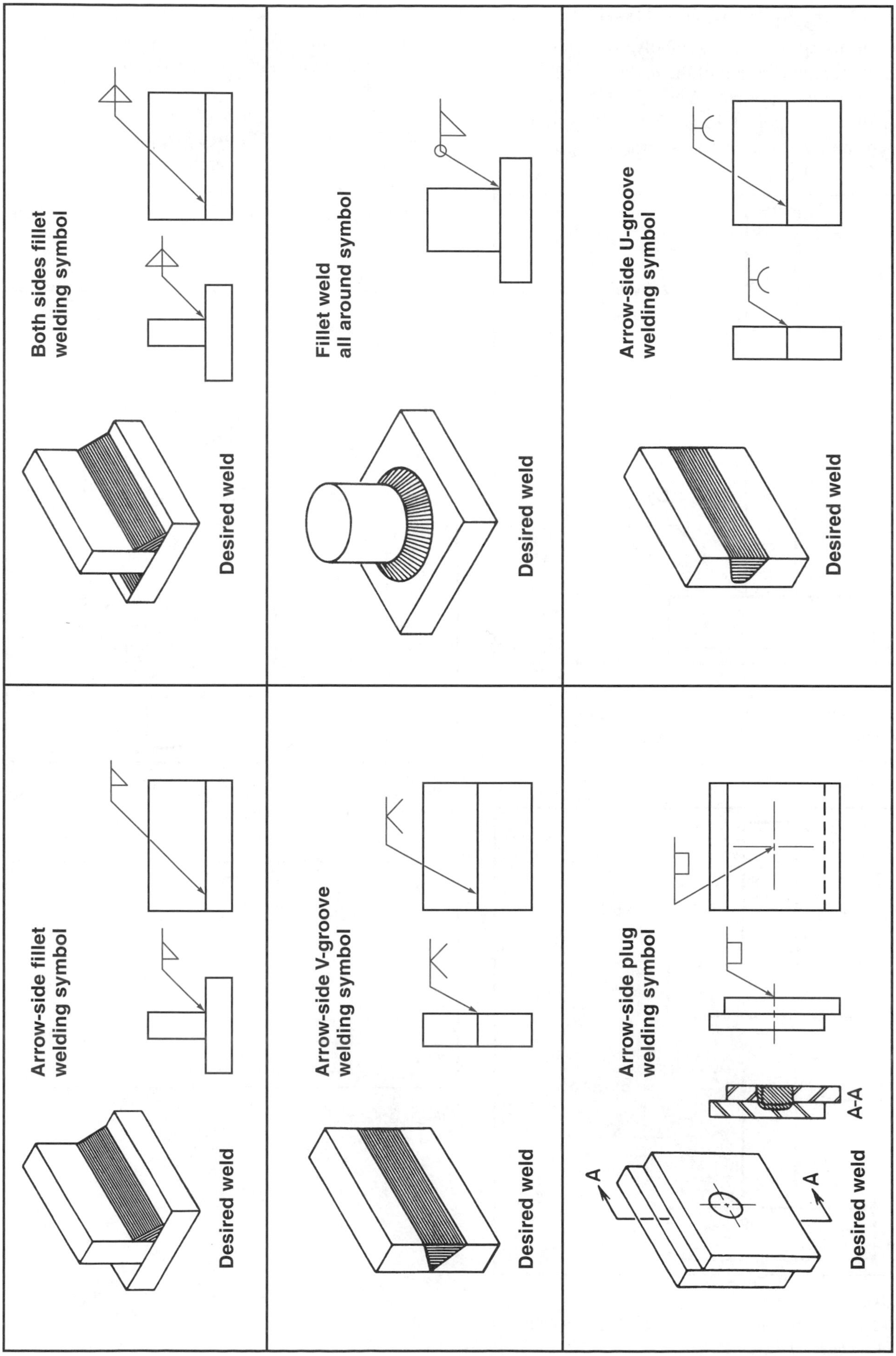

Both sides fillet
welding symbol

Desired weld

Fillet weld
all around symbol

Desired weld

Arrow-side U-groove
welding symbol

Desired weld

Arrow-side fillet
welding symbol

Desired weld

Arrow-side V-groove
welding symbol

Desired weld

Arrow-side plug
welding symbol

Desired weld

A-A

Figure 5-8.
Typical welding symbols and their meanings.

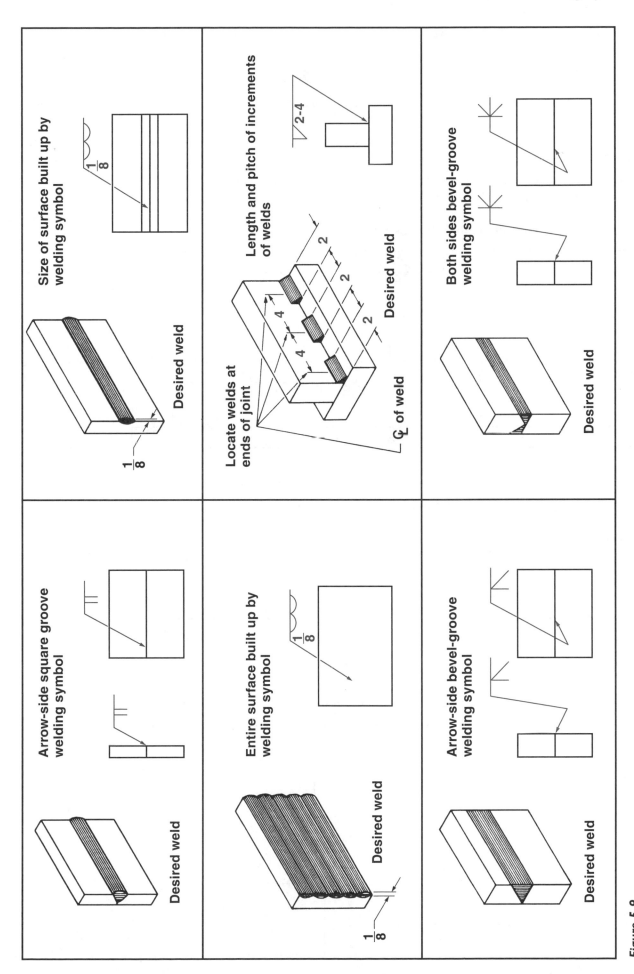

Figure 5-9.
Typical welding symbols and their meanings.

Name _____ Score _____

Things to Do

1. Ask your instructor for a metal part that contains a number of welds. Make a sketch of the part and show the correct welding symbol for each of the welds.

2. Design a metal part and make a drawing of each section that requires the use of weld symbols. Indicate the symbols on your drawing.

3. Using cardboard and glue, make models that show the use of bead, fillet, and plug welds. Place the correct weld symbol on each model with markers.

Problem 1: Draw the correct symbols for the following welds.

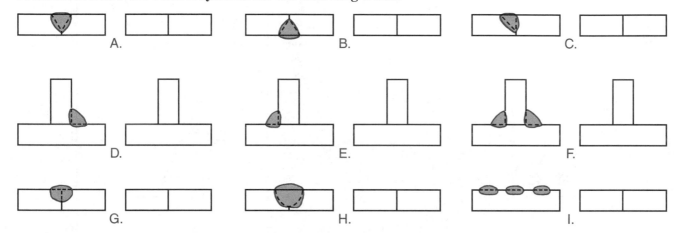

Problem 2: Draw the correct weld as indicated by the symbol.

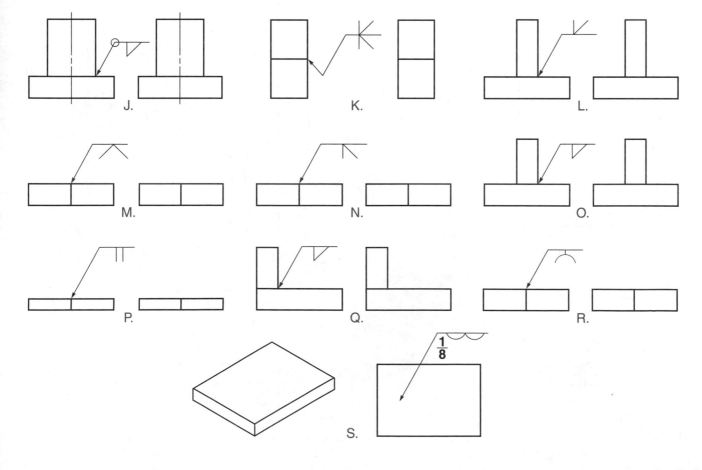

Unit 6
Arc Welding Equipment

After completing this unit, you will be able to:

○ Explain the difference between the two basic types of welding machines.

○ Identify the various pieces of equipment that complete the welding circuit.

○ List the basic weld cleaning and hot metal handling tools.

The equipment used for arc welding includes the power source, cables (leads), electrode holders, cleaning accessories, tools to handle hot metals, and protective clothing and devices.

Welding Machines

Welding machines, or power sources, are available in a large range of sizes and are rated by duty cycle. *Duty cycle* is the amount of time in a 10-minute period that the welding machine can be operated at a specified current, called rated current, without overheating. Duty cycle is given as a percentage at a specific current rating. For example, a machine with a duty cycle of 60% at 150 amps would be able to weld for 6 minutes at 150 amps without overheating. It must then be allowed to cool for at least 4 minutes before resuming the weld.

Welding machines are used to supply power to produce the arc, **Figure 6-1.** Some welding machines use power supplied by the power company; others are driven by an engine to produce power. The machines will generally provide Alternating Current (AC), Direct Current (DC), or both AC and DC. See **Figure 6-2.**

Figure 6-1.
Shielded metal arc welding (SMAW) being used in the field to construct a storage tank. (Miller Electric Mfg. Co.)

Figure 6-2.
AC arc welding machine. (Miller Electric Mfg. Co.)

Most welding machines used in industry for production and maintenance welding provide both AC and DC, **Figure 6-3.** Welding machines used for construction welding in the field are generally driven by a gasoline engine and produce DC for welding but will also produce AC to operate hand tools. Gasoline

engine-driven welders permit welding to be done in the field where there are no power lines, **Figure 6-4.**

Welders are also available that can be used to perform more than one welding process. Arc (stick) and gas tungsten arc welder combinations are common.

Figure 6-4.
Gasoline-driven AC-DC welder can be used where there are no power lines. (Miller Electric Mfg. Co.)

One example of this type of machine is an inverter. Inverter power sources are compact, lightweight welding machines, **Figure 6-5.** By modifying low input power to high frequencies, inverters achieve high-efficiency performance. These power

Figure 6-3.
AC-DC arc welding machine. (Miller Electric Mfg. Co.)

Figure 6-5.
This inverter arc welding machine uses electronics to produce efficient AC and DC power.

sources typically produce DCEN or DCEP. However, with additional equipment, inverters can also produce AC. Some inverters are also multi-process machines capable of shielded metal arc welding (SMAW), gas metal arc welding (GMAW), and sometimes gas tungsten arc welding (GTAW).

Leads

Two *leads,* or *cables,* carry current from the welding machine to the workpiece and back, **Figure 6-6.** These leads consist of heavy-duty, flexible, rubber-coated copper wire designed to endure the rough handling they receive in the welding shop.

An *electrode holder* is attached to one lead. The second lead is attached to the *ground clamp.* The clamp is attached to the work or to the welding table.

Lead size and length are determined by the amperage capacity of the welding machine.

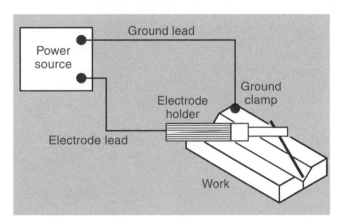

Figure 6-6.
The welding circuit.

Lugs, **Figure 6-7,** are fitted to the lead ends to provide a solid means of attaching the leads to the welder terminals and the ground terminal.

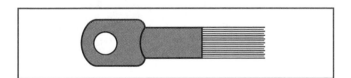

Figure 6-7.
Lug is fitted to end of cable to provide a solid means for attaching the lead to welding terminal.

Quick connects, **Figure 6-8,** allow the leads to be removed and replaced easily. To remove or attach a lead with a quick connect, align the screws on both sides of the connection. This will allow you to insert

Figure 6-8.
Quick connects are often used when leads are frequently removed from the welding machine.

or remove the lead. To secure the lead to the machine once the quick connect is inserted, turn the lead-side of the connection about one-half turn.

Electrode Holder

The *electrode holder,* **Figure 6-9,** must grip the electrode firmly and provide a good electrical contact. Its design should also permit quick and easy change of electrodes.

A holder with a comfortable handle with adequate heat insulation is recommended. For added protection, the outer portion of the jaws must be insulated to prevent grounding if the holder is accidentally touched to the work.

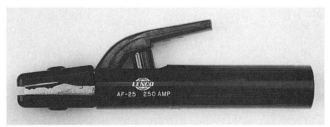

Figure 6-9.
Electrode holder. (Lenco-NLC, Inc.)

Ground Clamp

A *ground clamp,* **Figure 6-10,** provides the means for making a mechanically and electrically sound ground connection with the work or welding table.

A spring-pressure ground clamp is considered the most convenient type for general welding work.

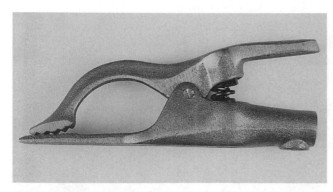

Figure 6-10.
Ground clamp. (Lenco-NLC, Inc.)

Weld Cleaning Equipment

A *wire brush* and *chipping hammer,* **Figure 6-11,** are the basic cleaning equipment used for removing slag and cleaning the weld bead.

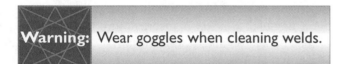

Warning: Wear goggles when cleaning welds.

Hot Metal Handling Tools

Conventional slip-joint pliers and tongs are recommended for handling work that has just been welded.

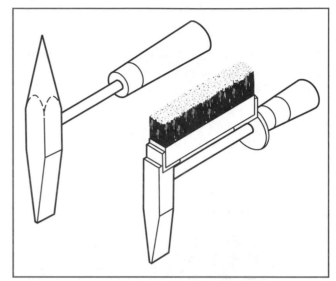

Figure 6-11.
Chipping hammers.

Name _____ Score _____

Check Your Progress

1. List the two basic types of arc welding machines.

   a. _____

   b. _____

2. When is a gasoline powered welder used?

   _____

   _____

3. Describe the cables used in arc welding.

   _____

   _____

4. _____ are fitted to cable ends to provide a solid means of attaching the leads to the welding machine terminals and the ground clamp.

5. The portions of the jaws of the _____ should be insulated to prevent _____.

6. The _____ and _____ are used to remove slag and clean the weld bead.

Things to Do

1. Visit a shop that does welding and observe the welding machines being used. Try to identify the types. Are they direct current, alternating current, or combination welders?

2. In the space provided, make a sketch showing a simple welding circuit. Label the various pieces of equipment that make up the circuit.

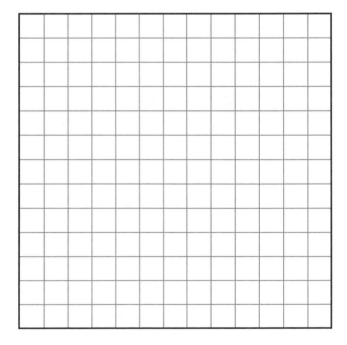

3. Write to manufacturers of arc welding equipment to secure their catalogs. Prepare a list of equipment with specifications and prices for a one-station arc welding area.

Notes

Unit 7

Types of Arc Welding Machines

After completing this unit, you will be able to:
- ○ Identify the common uses for both the direct-current, reverse-polarity machine setup and the direct-current, straight-polarity machine setup.
- ○ Describe the conditions that produce magnetic arc blow.
- ○ Create wiring diagrams for direct-current, reverse-polarity machine and direct-current, straight-polarity machine setup.

The current source in arc welding may be either *direct current (DC)* or *alternating current (AC)*.

DC Arc Welding Machines

The current flow on the DC welder is constantly in one direction and is expressed in terms of *polarity*. Electrodes are designed to take advantage of this condition.

When the positive (+) lead from the power source is connected to the electrode and the negative (−) lead is connected to the workpiece, the circuit is considered to be *direct current electrode positive (DCEP),* **Figure 7-1.** This is also known as a *direct-current, reverse-polarity (DCRP)* circuit. With this setup, the arc is forceful and digs into the base metal for deep penetration. It is used for most service welding.

When the negative (−) lead from the power source is connected to the electrode and the positive (+) lead is connected to the workpiece, the circuit is considered to be *direct current electrode negative (DCEN),* **Figure 7-2.** This is also known as a *direct-current, straight-polarity (DCSP)* circuit. The arc from this circuit is not as forceful and is used to weld thin material (sheet metal). The DCEN circuit also produces a rapid melt-off of the electrode, and metal is deposited about one-third faster than it is in the DCEP circuit.

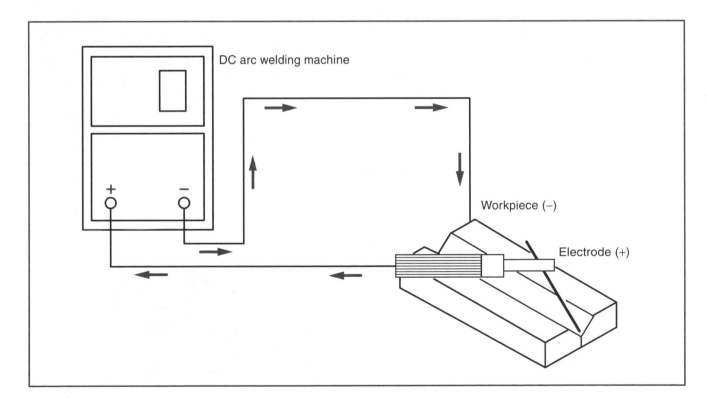

Figure 7-1.
A wiring diagram for direct current electrode positive (DCEP) arc welding circuit. It is also known as a direct-current, reverse-polarity (DCRP) circuit.

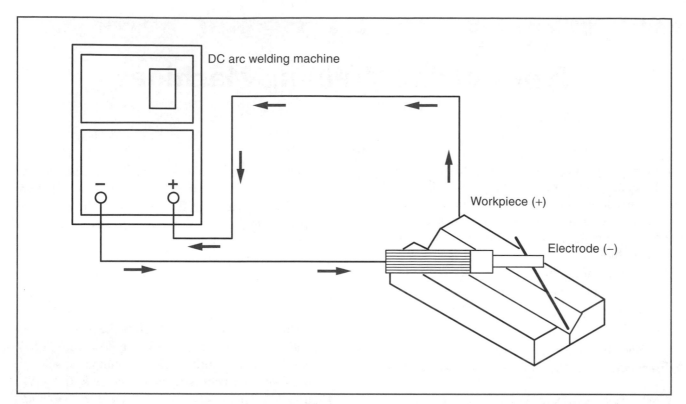

Figure 7-2.
The wiring diagram of a direct current electrode negative (DCEN) arc welding circuit. It is also known as a direct-current, straight-polarity (DCSP) circuit.

One condition that occurs when using DC welding machines is known as *magnetic arc blow.* Current passing through the work being welded establishes a magnetic field in the work, **Figure 7-3.** When these lines become concentrated (at the end of the joint or as the electrode nears a corner), they cause the arc to wander and a crater, or concave surface, forms.

An additional operation is often necessary to refill this depression. As the amperage increases, so does the magnetic arc blow problem.

The polarity used is determined by the type of electrode specified. Polarity may be changed by a switch on the welder.

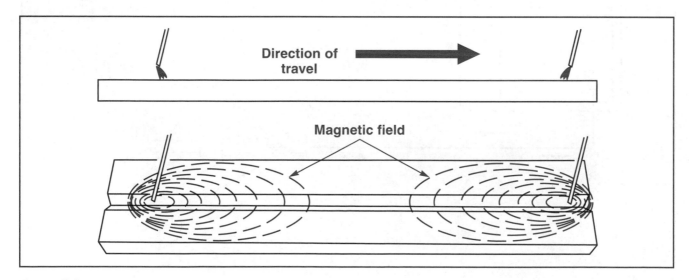

Figure 7-3.
The magnetic field that causes magnetic arc blow is generated when doing DC welding. Note how the arc changes as it moves toward the end of the weld.

AC Arc Welding Machines

In the AC welder, the current reverses its direction of flow 120 times a second. Alternating current in the U.S. operates at 60 cycles, which means that it goes through a complete cycle, or a change in direction of current flow from positive to negative and back to positive, 60 times a second. This reversal almost completely eliminates the problem of magnetic arc blow because it keeps the effect of the magnetic field on the arc to a minimum.

The agitation caused by the reversal of the current flow causes the metal in the electrode and elements in the electrode covering to mix thoroughly with the metal of the work and produce a uniform, dense weld. The pulsating effect on the molten metal also tends to cause the slag and impurities to come to the weld surface, where they can be chipped away.

AC/DC Welding Machines

Most machines used for production, construction, and maintenance welding produce both AC and DC straight or reverse polarity. DC is produced by passing AC from the power supply through a semiconductor circuit called a *rectifier*. A switch on the welding machine will let you select AC, DCEP, or DCEN.

Robot welders perform repetitious tasks on vehicles being assembled in a modern car manufacturing plant. (Ford Motor Company)

Name _____ Score _____

Check Your Progress

1. *Polarity* is determined by: (Check correct answer.)
   a. _____ electrode to be used.
   b. _____ type of current available.
   c. _____ size of the welder.
   d. _____ All of the above.
   e. _____ None of the above.

2. What problem is caused by *magnetic arc blow?*

   _____

   _____

3. The current source in arc welding may be either _____ _____ or _____.

4. The current produced by the DC welder flows constantly in one _____.

5. Current passing through the work being welded establishes a _____ _____ in the work.

6. The pulsating effect of some welders on the molten metal tends to cause _____ and _____ to come to the surface.

Things to Do

1. In the space provided, make a sketch of a *direct-current, straight-polarity* circuit. Label the various parts of the circuit.

2. In the space provided, make a sketch of a *direct-current, reverse-polarity* circuit. Label the various parts of the circuit.

3. Prepare a bulletin board display illustrating the various types of arc welding machines. Use clippings from manufacturers' catalogs and technical magazines. Indicate the type of welder, polarity, lead connections, and type of work for which each is generally used.

4. Make a sketch of a DC arc welding machine from your shop. Label all the parts and describe the use of each.

Unit 8
Electrodes

After completing this unit, you will be able to:
- ○ Explain the purpose of the flux coating on electrodes.
- ○ Define the significance of each position in the classification number for welding electrodes.
- ○ Describe the proper storage of electrodes.

Electrodes support the electric arc and provide filler metal to the joint. They are covered with a flux coating that is melted by the high temperature generated by the arc. As it melts, the flux coating:
- • Cleans the surface being welded.
- • Releases gases that shield the molten weld area from being contaminated by oxygen and nitrogen in the atmosphere. The atmospheric gases weaken the weld, **Figure 8-1.** The gas also helps to conduct electricity.
- • Produces a slag that acts as insulation and slows the rate of cooling of the weld. This reduces internal strains that develop from sudden changes in temperature.
- • Reduces spatter.
- • Makes easier starting of the arc.
- • Helps ensure even distribution of the metal in the rod with the parent metal.
- • Adds alloy materials to the weld metal.

Classifying Electrodes

Electrodes used in arc welding can be classified according to their operating characteristics, type of coating, and qualities of the deposited metal. They can be further subdivided according to coating composition, welding current, and welding position. See **Figure 8-2.**

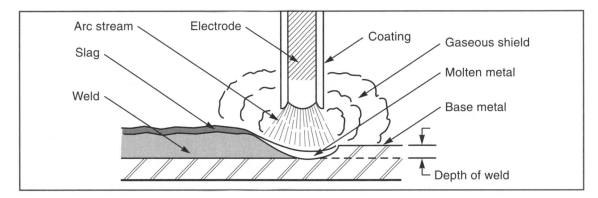

Figure 8-1.
Close-up of welding operation. Note how the gases released by the coating cover the weld area.

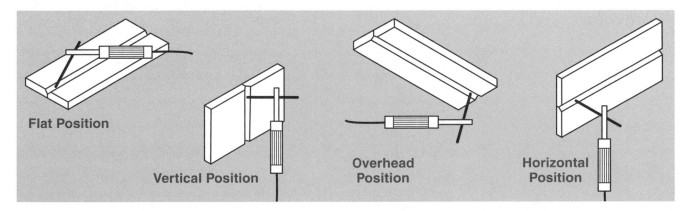

Figure 8-2.
The four welding positions.

To end the confusing task of selecting the correct electrode for a specific job, the American Welding Society (AWS) has established a uniform numbering system for classifying electrodes.

The AWS classification number is stamped on the coating near the grip end of the electrode, **Figure 8-3.**

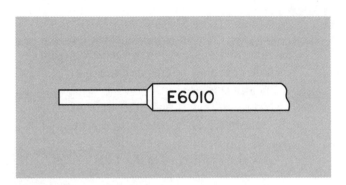

Figure 8-3.
Location of the AWS numbering system.

Explanation of AWS Classification Numbers (Steel Arc Welding Electrodes)

All mild steel and low alloy electrodes are classified with a four- or five-digit number prefixed by the letter "E."

Prefix "E" = Electrode for arc welding.

First two (or three) digits = Tensile strength (Tensile strength is the resistance of a material to forces trying to pull it apart. It is given in pounds per square inch [psi]).

Third (or fourth) digit = Position of welding.

1 = All positions (flat, horizontal, vertical, overhead).

2 = Horizontal and flat positions only.

3 = Flat position only.

Fourth (or fifth) digit = Type of coating and welding current.

Fourth Digit	Type of Coating	Welding Current
0	Cellulose sodium	DC Reverse
1	Cellulose potassium	AC or DC Reverse
2	Titania sodium	AC or DC Straight
3	Titania potassium	AC or DC Straight
4	Iron powder titania	AC or DC
5	Low hydrogen sodium	DC Reverse
6	Low hydrogen potassium	AC or DC Reverse
7	Iron powder iron oxide	AC or DC
8	Iron powder low hydrogen	AC or DC Reverse

For example, see **Figure 8-4.**

When the fourth digit is "0," the type of power supply is determined by the third digit.

A list of commonly used electrodes is given in **Figure 8-5.** For high-strength electrodes and alloy electrodes, a suffix to the electrode number is used. The suffix appears after a dash and indicates the specific chemical additions to the electrode. For example, E7018-B2 electrode has chromium and molybdenum added to it.

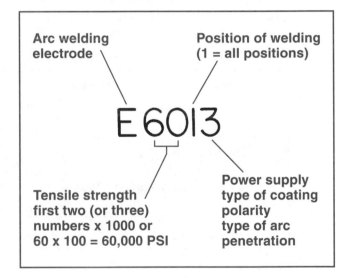

Figure 8-4.
The AWS numbers are used to identify the characteristics of an electrode.

Welding Characteristics of Mild Steel Electrodes						
Classification	Welding Current	Welding Position	Penetration	Slag	Spatter	Application
E6010	DC Reverse	All	Deep	Thin	Moderate	Best all position electrode
E6011	AC–DC	All	Deep	Thin	Moderate	Same as E6010 except primarily for AC current
E6012	AC–DC Straight	All	Medium	Heavy	Slight	For joints with poor fit
E6013	AC–DC Reverse or straight	All	Shallow	Medium	Slight	Best general purpose electrode for mild steel
E6024	AC–DC Straight	Flat horizontal	Shallow	Heavy	Slight	Best where high deposition rate is required
E7018	AC–DC Reverse	All	Medium	Medium	Very slight	For low and medium carbon and low alloy steel
E11018	AC–DC Reverse	All	Medium	Medium	Slight	For heat treated steel

Figure 8-5.
Commonly used electrodes.

Storing Electrodes

Electrodes must be kept dry. Moisture destroys the desirable characteristics of the coating. It may cause excessive spattering and lead to the formation of cracks in the welded area. Low hydrogen electrodes (EXXX5, 6, 8) should be stored in an electrode oven.

Electrodes exposed to damp air for more than two or three hours should be dried by heating in a suitable oven for two hours at 500°F (260°C). After they have been dried, the electrodes should be stored in a suitable container.

Name _____ Score _____

Check Your Progress

1. Identify the various parts of the illustration shown below.

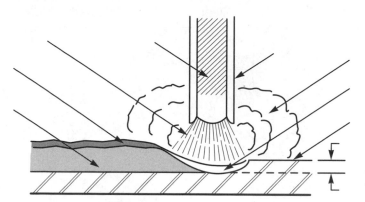

2. The coating on the electrode aids in making the weld by: Check correct answer(s).

   a. _____ keeping the electrode from rusting.

   b. _____ reducing spatter.

   c. _____ keeping the weld from cooling too rapidly.

   d. _____ preventing contamination of the weld by oxygen and nitrogen in the atmosphere.

   e. _____ ensuring even distribution of the metal in the electrode and parent metal in the weld.

3. What does AWS stand for?

   _____

4. Sketch and label the four (4) positions of welding.

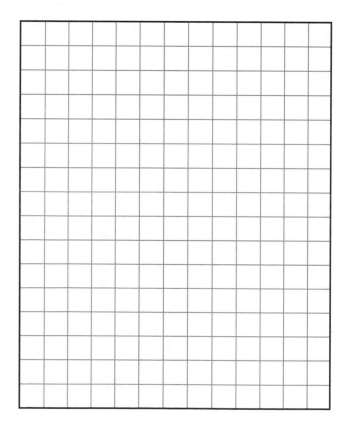

5. Using the information in **Figure 8-5,** list the electrodes best suited for the following jobs:

   a. For joints with poor fits. _____

   b. Best all-position electrode. _____

   c. For heat-treated steels. _____

   d. For medium carbon and low alloy steels. _____

   e. Best general purpose electrode for mild steels. _____

6. What welding problems can be caused by using electrodes that have absorbed moisture?

   _____

   _____

7. AWS classifies electrodes by: Check correct answer(s).

   a. _____ diameter.

   b. _____ color of the electrode coating.

   c. _____ a uniform numbering system.

   d. _____ All of the above.

   e. _____ None of the above.

Things to Do

1. Secure several different types of welding electrodes. Note how they are marked, the thickness of the coatings, the characteristics of the coatings, the colors of the coatings, and the sizes (diameters) of the rods.

2. Examine the containers that held the electrodes when they were purchased. How did they keep the electrodes moisture free? What information was on the container concerning the electrodes?

3. Look over welds made with different types of electrodes. Identify the rods used and note the difference in weld appearance.

Notes

Unit 9
Selecting the Proper Electrode

After completing this unit, you will be able to:
○ List the factors to consider when choosing an electrode for a specific job.
○ Explain how the weldability of steel is determined.
○ Summarize the general rule for selecting the appropriate electrode size for a weld.

One of the most important factors in making a sound weld is the selection of the proper electrode for the job. Points to be considered in choosing the electrode include:
1. Type of metal being welded.
2. Thickness of metal being welded.
3. Position of the joint.
4. How rapidly the electrode metal is to be deposited in the weld.
5. Type of joint preparation.
6. Welding code requirements.

The weldability of steel is determined by its carbon or alloy content. The chart on page 47 will prove helpful in determining the proper electrode to use. For more complete information, refer to the charts prepared by electrode manufacturers.

To select the correct size (diameter) electrode for the weld to be made, a good rule is never use an electrode with a diameter larger than the thickness of the metal to be welded. See **Figure 9-1.** However, it is advisable to use as large an electrode as possible. This will permit increased welding speeds and lower costs.

When making fillet welds, it is generally recommended that you use an electrode slightly smaller than the specified fillet, **Figure 9-2.** Overhead and vertical welding can be done with a minimum of effort using 1/8″ or 5/32″ (3 mm to 4 mm) diameter electrodes.

Thick metal requiring a multiple pass groove weld is best done by utilizing a 1/8″ (3 mm) diameter electrode to make the first two passes. This will ensure full penetration. Additional passes are then made with larger diameter electrodes.

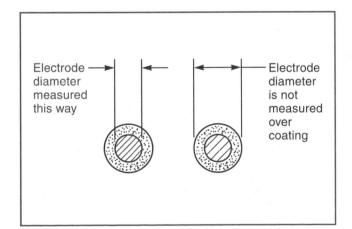

Electrode diameter measured this way

Electrode diameter is not measured over coating

Figure 9-1.
Measure the electrodes as shown at left. The coating is *not* included in the diameter.

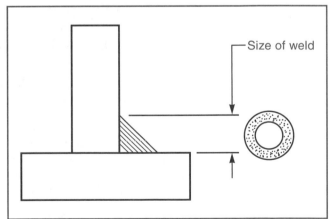

Size of weld

Figure 9-2.
When making fillet welds, select an electrode of slightly smaller diameter than the size of the desired fillet.

Name _____ Score _____

Check Your Progress

1. List five factors that should be considered when choosing electrodes for a specific job.

   a. _____

   b. _____

   c. _____

   d. _____

   e. _____

2. The weldability of steel is determined by the: Check the correct answer(s).

   a. _____ shape of the joint.

   b. _____ thickness of the metal.

   c. _____ carbon content.

   d. _____ alloying elements.

3. To select the correct size electrode for the job, a good rule to follow is _____

   _____

   _____

   _____ .

4. In general, when making fillet welds, it is recommended that you use an electrode that is _____ than the specified fillet.

5. Best results in making overhead and vertical welds are obtained using _____ or _____ diameter electrodes.

   _____

   _____

6. Solid welds in thick metal can best be made if _____

   _____ .

Things to Do

1. From reference materials and manufacturer's literature, prepare a chart listing eight different types of steel and the appropriate electrode best suited to weld each type. Include steels from low carbon content to high alloy steels.

2. Prepare a written report on specialized arc welding electrodes available to the welder. Be sure to include the importance of selecting any specialized electrodes for their intended use.

3. Make a list of the things that may happen to a weld, or during welding, if an improper electrode is used.

Unit 10
Joint Preparation

After completing this unit, you will be able to:
○ Cite the danger created by a weld that does not penetrate close to 100%.
○ Explain why certain joints are recommended for materials of specified thickness.
○ List the precautions that should be observed before starting a weld.

The next step in the welding sequence is to select the type of joint to be used. In general, the best joint is the simplest joint that will adequately do the job required.

A weld must penetrate close to 100%; otherwise, it will be weaker than the base metal. Get the deepest penetration possible.

A square butt joint, **Figure 10-1,** may be used on material 1/4″ (6 mm) or less thick. This joint can be used because there is enough penetration in one welding pass to assure a good weld.

The single bevel groove or single V groove joint is recommended for material 1/4″ to 3/4″ (6 mm to 19 mm) thick. For complete penetration, the joint should be prepared as shown in **Figure 10-2.** A back bead should be made on this type of joint to reduce the possibility of warping.

The joints are made by grinding, machining, cutting with an oxyacetylene torch, or cutting with a plasma torch.

A double bevel or double V groove joint is frequently specified on thicker material instead of the single bevel or single V groove joint. With double joints, warping is minimized because the metal is deposited equally on both sides of the joint and not as much welding time and electrode material is required, **Figure 10-3.**

Often, J and U groove joints are specified on material 3/4″ (19 mm) or thicker, **Figure 10-4.** This is for economic reasons, because they require less welding time and electrode material. Again, a double J or U groove joint is preferred over the single groove joint.

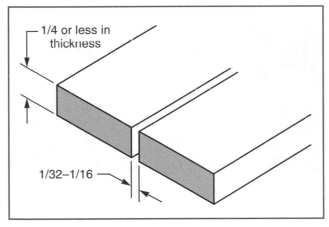

Figure 10-1.
A square butt joint is satisfactory when welding material 1/4″ thick or thinner.

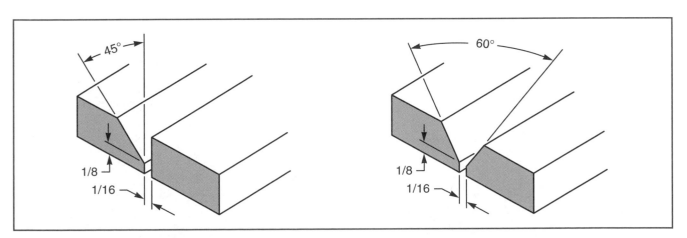

Figure 10-2.
Single bevel groove and single V groove joints.

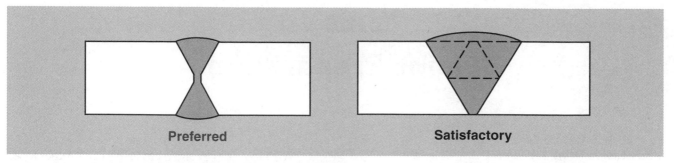

Preferred   Satisfactory

Figure 10-3.
A double V groove joint is preferred over the double bevel groove joint because it is less expensive to weld.

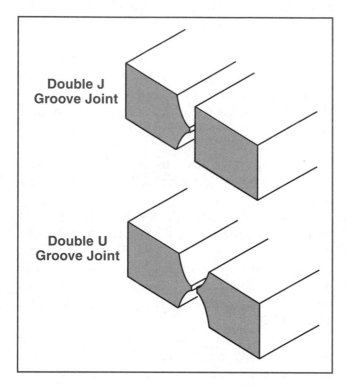

Double J
Groove Joint

Double U
Groove Joint

Figure 10-4.
Double J and double U groove joints.

Clean/Dry Joints

Before attempting to weld, remove all dirt, rust, grease, or other material from the joint. Leaving these foreign materials on the joint will usually result in a weld with greatly reduced strength.

If dirt is removed with flammable solvents, be sure the joint is thoroughly dry and remove solvent containers from the area before attempting to weld.

Warning: Use extreme care when using an oxyacetylene torch to remove paint and other coatings from the weld area. The fumes generated may be dangerous to your health and could cause long-term illness.

Name _____ Score _____

Check Your Progress

1. In general, the best joint is the _____

    _____

    _____.

2. The weld must penetrate close to 100%; otherwise: Check the correct answer(s).
    a. _____ the weld will be too strong.
    b. _____ the weld will not be as strong as the base metal.
    c. _____ the weld will not be complete.
    d. _____ All of the above.
    e. _____ None of the above.

3. A(n) _____ joint may be used when the material is 1/4″ (6 mm) thick or less.

4. Make a sketch of the above joint.

5. The _____ or _____ groove joints are recommended when the material is 1/4″ to 3/4″ (6 mm to 19 mm) thick.

6. Make sketches of the above joints.

7. When making welds in material thicker than 3/4″, double J and double U groove joints are preferred over single groove joints because: Check correct answer(s).
    a. _____ they are more economical.
    b. _____ less time is required to complete the welds.
    c. _____ less welding electrode material is needed.
    d. _____ warping is minimized.

8. What precautions (safety and otherwise) should be taken before starting to weld?

a. _____

b. _____

c. _____

Things to Do

1. Make sketches of double J and U groove joints.

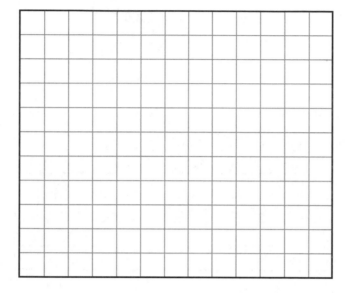

2. Complete the following drawings of joint preparation using the welding symbols as a guide.

A

B

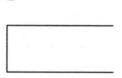

C

D

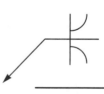

E

Unit 11
Preparing to Weld

After completing this unit, you will be able to:
- List the two methods for striking the arc.
- Demonstrate the electrode position when running a flat weld.
- Determine the necessary adjustments to the current, arc length, or speed by examining test welds.

The first step in preparing to weld is to carefully check over the machine. Be sure all connections are clean and tight. Since the heat generated by the arc is determined by the amount of current (amperage) used, set the machine to the range recommended for the size electrode being used.

If the manufacturer's recommendations are not available, a good rule of thumb when using standard electrodes is to adjust the welding machine to a current setting that is approximately equal to the diameter of the electrode in thousandths of an inch. For example, a 1/8″ (0.125″ or 3 mm) diameter rod will operate at ± 125 amperes. (This may vary slightly for electrodes made by different manufacturers.)

When you begin to weld, a few trail beads should be completed. Raise or lower the welding machine current setting until you get a satisfactory weld. A weld should not burn through, but it should have sufficient penetration, **Figure 11-1.**

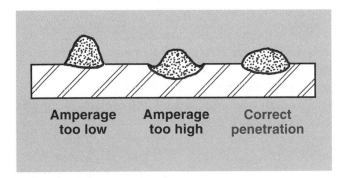

Figure 11-1.
Weld characteristics.

Position the metal, securing it with clamps if needed. Attach the ground clamp securely. Clamp the electrode in the electrode holder at a 90° angle to the holder jaws, **Figure 11-2.**

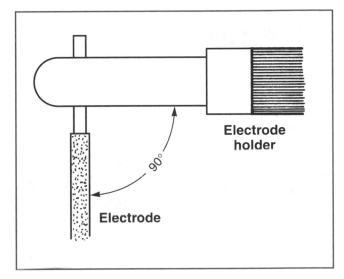

Figure 11-2.
Electrode position in electrode holder for running a flat weld.

Warning: Put on your protective clothing before welding. Replace any items that are not up to standards. Remove all flammable materials and solvents from the welding area and put shields in place if others will be working in the immediate area.

Keep the electrode and holder clear of the work area and turn on the machine. Grasp the electrode holder with a comfortable grip (do not hold it too tightly or your hands will tire quickly). Place the clamp section of the electrode holder toward your thumb. This will give you extra leverage for striking the arc and moving the electrode. Use two hands whenever possible, **Figure 11-3.** Lower the electrode to about 1″ (25 mm) above the work. Lean the electrode to a 15–20° angle from the vertical in the direction of travel, **Figure 11-4.** The inclination is important. If the angle is more than 20°, the penetration may not be enough to produce a solid weld.

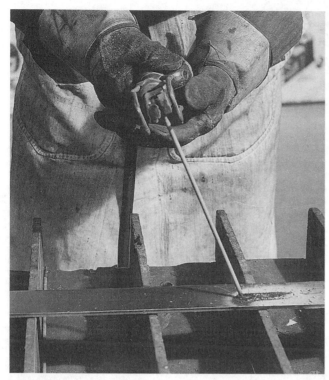

Figure 11-3.
Use two hands to grasp the electrode holder whenever possible. Use a comfortable grip or your hand(s) will tire rapidly. (Lincoln Electric Co.)

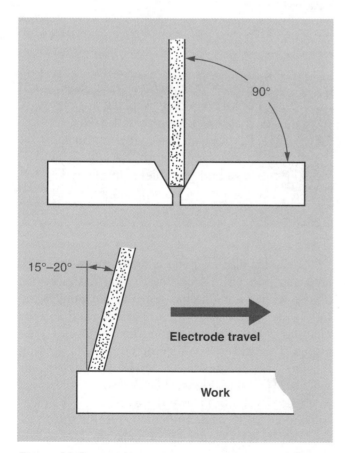

Figure 11-4.
Electrode position when running a flat weld.

Striking the Arc

Lower your head shield and strike the arc. There are two methods for striking the arc, the scratch method and the tap method, **Figure 11-5.** The scratch method is recommended for beginning welders. The tap method is sometimes used after the welder gets some practice.

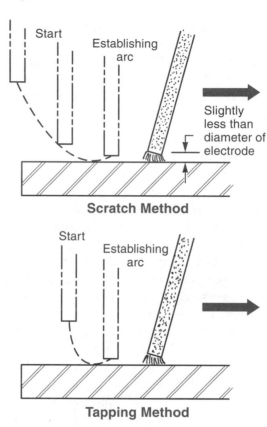

Figure 11-5.
Methods for starting the arc.

To strike an arc using the *scratch method,* scratch the tip of the electrode over the work much like you would strike a match. As the arc starts, hold a slightly longer than normal arc until the base metal is heated enough to form a proper size pool of molten metal. Keep the electrode moving or it will "freeze," or stick to the work.

To strike an arc using the *tap method,* bring the electrode close to the location where the weld is to begin or continue. Lightly tap the electrode on the work. When the arc starts, hold a slightly longer-than-normal arc for about one second and then establish the proper arc length. It will require some practice to master the force required to start the arc by the tap method. If you apply too much force, the electrode will stick to the work. Without enough force, the arc will not start. The tap method requires less area to strike the arc and will enable you to strike the arc without damaging the base metal around the joint.

If the electrode should weld fast (stick) to the work, break it loose by twisting or bending the electrode by sharply moving the holder. If this does not work, squeeze the clamp section of the electrode holder and release the electrode from the holder.

Once the length of the arc is established, try to maintain the correct length as the electrode burns down. If the arc is too short, not enough heat will be generated to melt the parent metal and the rod may stick. If the arc is too long, excessive spattering may occur and an irregular bead with poor penetration will result.

In general, use an arc length equal to or slightly less than the core diameter of the electrode. Since the arc length cannot be measured, practice running beads until you can tell when you are maintaining the correct arc length and welding speed. Many welders go by the sound of the arc to help them determine correct arc length.

Maintaining the correct welding speed is important. Watch the pool of molten metal directly behind the arc. Do not watch the arc. The shape of the weld pool and the ridges formed when the molten metal solidifies indicate correct welding speed. See **Figure 11-6.** The ridges should be uniform and straight.

Common mistakes made when learning to weld are shown in **Figure 11-7.** Study them carefully so you can identify mistakes in your welding.

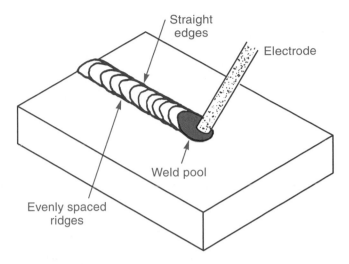

Figure 11-6.
The completed arc bead should have straight edges, evenly spaced ridges, and uniform height.

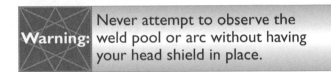

Warning: Never attempt to observe the weld pool or arc without having your head shield in place.

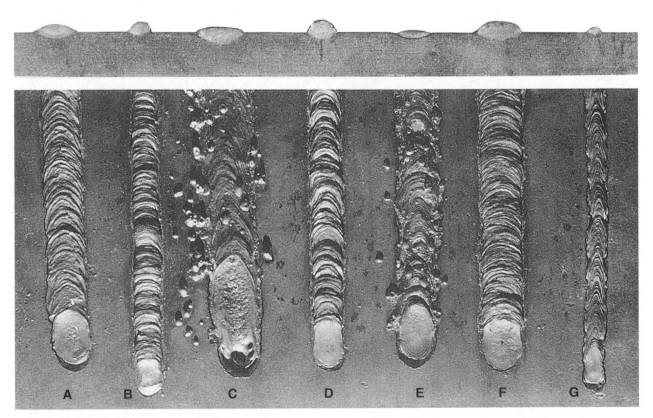

Figure 11-7.
Characteristics of welds made under various conditions. The conditions are accentuated to illustrate differences. A—Current, voltage, and speed normal. B—Current too low. C—Current too high. D—Arc too short. E—Arc too long. F—Speed too low. G—Speed too high. (Lincoln Electric Co.)

Name _____ Score _____

Check Your Progress

1. List three safety precautions that should be observed when preparing to weld.

    a. _____

    b. _____

    c. _____

2. List the two methods prescribed for striking an arc:

    a. _____

    b. _____

3. How can an electrode that has become welded fast to the work be freed?

    _____

    _____

4. Prepare a sketch showing the position of the electrode for flat welding.

5. What is the general rule for establishing the correct length of the arc?

    _____

    _____

6. In general, the current setting for an electrode is approximately equal to the diameter of the electrode in thousandths of an inch. If this is true, what are the current settings for the following electrodes?

    a. 1/8″ (3 mm) _____

    b. 5/32″ (4 mm) _____

    c. 3/16″ (5 mm) _____

    d. 1/4″ (6 mm) _____

Things to Do

1. Have your instructor or an experienced welder demonstrate the proper way to do the following:
   a. Check out the welding machine and related equipment.
   b. Inspect your welding shield and protective clothing.
   c. Adjust the machine(s) to get the correct setting and attach the ground cable.
   d. Select the proper electrode for the job.
   e. Install the electrode in the holder.
   f. Strike and maintain the arc.
   g. Break free an electrode that has become welded to the work.

2. Have your instructor or an experienced welder run several beads under various conditions, such as current too low, current too high, arc too short, arc too long, speed too low, speed too high, etc.

3. Practice checking over the welding machine.

4. Practice checking over your shield and protective clothing.

5. Practice inserting an electrode in the holder.

6. Practice (with the machine *off*) striking an arc.

Notes

Unit 12
Running Short Beads

After completing this unit, you will be able to:

○ Thoroughly prepare the work material for a practice weld.

○ Run a short bead with uniform ripples and little spatter.

○ Clean the weld with a chipping hammer and wire brush.

To develop the skills needed to become a good arc welder, you must perform certain prescribed exercises. This exercise, *Running Short Beads,* will introduce you to your first arc welding task, striking and maintaining an arc. Carefully study **Figure 12-1** and secure the necessary material. Wear your safety goggles or safety glasses and remove all burrs and sharp edges from the work with a file or grinder. Use a 1/8″ (3 mm) diameter E6012 or E6013 electrode. Use chalk or soapstone to draw lines on your practice piece as shown in Figure 12-1.

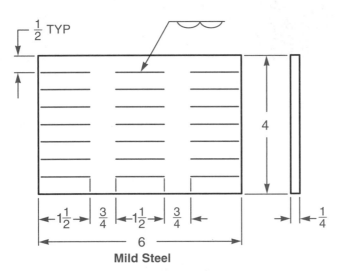

Mild Steel

Figure 12-1.
Specifications for this practice piece.

Procedure:

Note: Fill out the *Check List* at the end of this unit as you complete the following procedure.

1. Check your safety equipment and put it on. Also, be sure that other safety items, such as flash shields or curtains, are in place.

2. Position the work and attach the ground clamp. On work this size, it may be necessary to attach the metal to the welding table or bench.

3. Put the bare end of the electrode in the electrode holder, **Figure 12-2.** When doing this, keep the holder and electrode clear of the work area. When the holder is not in use, hang it in the place provided.

4. Turn the machine *on* and adjust it to the correct setting for polarity (if a DC machine is used) and amperage. For example, the E6012 electrode is designed for *straight* polarity. Connect the negative (−) lead to the electrode. The amperage will be approximately equal to the diameter of the electrode in thousandths of an inch. A 1/8″ (3 mm) electrode equals 0.125″, so the starting point for the amperage setting should be 125 amps. If, after you have started to weld, you find that the arc is too hot (excessive spattering with an irregular deposit of metal), reduce the amperage setting. On the other hand, trying to weld with a current that is too low will result in a weld with poor penetration or an electrode that sticks or "freezes" to the work. This happens because the arc does not generate enough heat to melt the parent metal. Make adjustments in 5 amp increments.

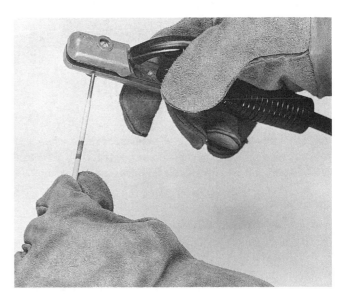

Figure 12-2.
Inserting electrode in electrode holder.

5. Assume a comfortable position and grasp the holder in a relaxed manner. If you grip it too tightly, you will tire quickly.

6. Place the electrode in position about 1" (25 mm) above starting point, lower your helmet, and strike the arc. You may use one of the techniques described in Unit 11. The scratch method is recommended.

7. Run a bead over a few of the lines on your practice piece. Move from left to right if you are right-handed (or from right to left if you are left-handed) with the electrode slightly tilted in the direction of travel. As the arc starts, hold a slightly longer-than-normal arc until the base metal is heated enough to form a proper size weld pool. Keep the electrode moving or it will "freeze," or stick to the work.

   a. The correct arc length produces a steady frying and crackling sound.

   b. There is little spatter and the ripples produced on the bead will be uniform and smooth, with no overlap or undercut, **Figure 12-3.**

   c. Remove the electrode from the holder when it is used to about 1 1/2" to 2" long. The electrode is hot. Put it in a location where you will not touch or step on it.

8. With safety glasses or goggles in place, remove the slag and examine the weld. Compare the weld bead with the weld characteristics shown in Figure 11-7. Use extreme care, as the metal will be hot.

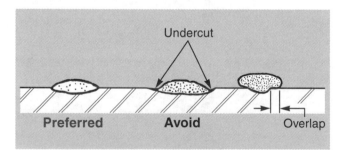

Figure 12-3.
You are following the correct welding procedure when there is little spatter and the ripples produced on the bead are uniform and smooth, with no overlap or undercut.

9. Make changes in the machine settings, if necessary, and run another bead. Follow this procedure until you produce a satisfactory bead.

Cleaning the Weld

When you have finished running a bead over each line on your practice piece, clean the welds completely. Remove the slag by striking it with the chipping hammer. Hit the bead in such a way that the slag flies away from you. Run the point of the hammer along the edges of the weld to remove the remaining slag. Clean the weld thoroughly with a wire brush.

You can shorten the waiting time for examining the completed weld and reduce the danger from flying slag when you clean the weld by first dipping the hot metal in cold water. This procedure should be used on exercise and practice welds only, since sudden cooling may affect the integrity (soundness) of the weld.

Examine all the weld beads and compare them to the characteristics shown in Figure 11-7. If you leave your welding machine, make sure to remove the electrode from the holder. Otherwise, the electrode could touch the machine or table and begin to arc while you are away from the machine.

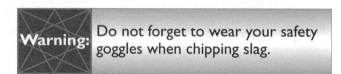

Warning: Do not forget to wear your safety goggles when chipping slag.

Safety Hints

1. When you are not using your electrode holder, hang it in the place provided. Do not lay it on the bench when the machine is on.

2. Turn the welder off when leaving the welding table.

3. Check over the welding area when you are finished welding. Turn machine off. Hang equipment in the places provided.

4. Do not get oil or grease on your safety clothing.

5. Remove paint, dirt, or grease from your practice piece before attempting to weld on it.

Name _____   Score _____

Check Your Progress

1. Remove the slag by striking it with the _____ hammer.

2. When not using the electrode holder:
   a. _____ Lay it in a safe place on the floor.
   b. _____ Hang it on the bench, ready for further welding.
   c. _____ Hang it in the place provided in the booth.
   d. _____ Lay it on a piece of nonmetallic material.

3. List four ways you can determine if your welding procedure is correct.
   a. _____
   b. _____
   c. _____
   d. _____

4. Welding with a current that is too low will result in a weld with poor _____.

5. Keep the electrode moving as you are welding or it will _____ to the work.

Things to Do

Before starting to weld, fill out the following *Check List.*

1. List of tools and equipment needed:
   _____
   _____

2. Check out safety clothing and equipment.
   Helmet _____
   Gloves _____
   Jacket _____
   Apron _____
   Trousers _____
   Shoes _____
   Safety goggles _____
   Flash shields _____

3. Check over welding area.
   Properly ventilated _____
   Flammable materials removed _____
   No moisture in area _____
   No solvents or thinners in area _____

4. Check over equipment.

Machine _____

Holder _____

Ground clamp _____

Cables _____

Table _____

Ventilating equipment operating _____

5. Burrs and sharp edges removed from work?

6. Electrode information.

Size specified _____

Type _____

Coating solid _____

7. Machine settings.

Polarity _____

Amperage _____

After you have finished the exercise, complete this portion of the *Check List*.

Problem	Cause	Correction
Example: Electrode "freezes" to work.	Inexperience	Practice

Unit 13
Running Continuous Beads

After completing this unit, you will be able to:

○ Run a continuous bead that is uniform in shape and size and shows ample penetration.

○ Stop the bead at the end of a weld by correctly backing the electrode to break the arc.

○ Restart the arc on an existing weld to complete the bead.

This welding exercise is designed to give you experience running weld beads that are straight and uniform in size and appearance. Much practice is necessary before you will be able to run a bead that is up to acceptable standards. You will also encounter the problem of restarting an arc after it is accidentally extinguished or after the electrode has been changed.

Study the plan in **Figure 13-1** carefully, and secure the necessary material. Use 1/8" (3 mm) diameter E6012, E6013, or E7018 electrodes. Do not discard electrodes until they are 1 1/2"–2" (38 mm–51 mm) long. They are expensive. Use them wisely.

Procedure:

1. Put on your safety glasses and cut or shear your metal to size. Carefully remove all burrs and sharp edges.

2. Use soapstone or chalk to draw a series of lines on the steel plate, **Figure 13-2.**

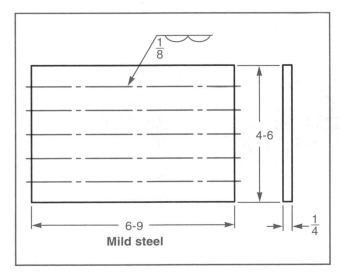

Figure 13-1.
Plans for an exercise in running a continuous bead.

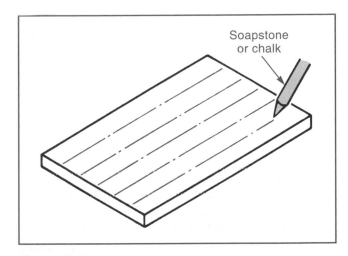

Figure 13-2.
Use soapstone or chalk to draw the lines on the practice metal. Use both sides of the practice piece to conserve material.

3. Check your safety equipment and put it on. Turn the welding machine to *on* and adjust it to the correct settings for polarity and amperage.

4. Position the work and attach the ground clamp.

5. Place an electrode in the holder and strike an arc. If you are right-handed, run a continuous bead starting at the left edge of the work. If you are left-handed, run a continuous bead starting at the right edge of the work. This position and direction will be most comfortable for you and give you the best results. Try to move the electrode from side to side slightly (no more than two electrode diameters). This will keep the puddle fluid, help release gas from the puddle, and make the slag easier to remove. Movement patterns are shown in Unit 15. At the end of the weld, stop the bead by shortening the arc and quickly backing, or "whipping," the electrode to break the arc, **Figure 13-3.**

6. With your safety glasses in place, remove the slag and examine the weld. Compare the weld bead with the weld characteristics shown in **Figure 11-7.** Use extreme care, as the metal will be hot.

7. Make changes in the machine settings as necessary and run another bead. Continue this procedure until you can produce a bead to acceptable standards.

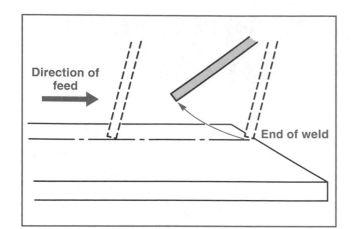

Figure 13-3.
At the end of the weld, stop the bead by shortening the arc and quickly backing, or "whipping," the electrode to break the arc.

Should you have to stop the bead for any reason, clean the end of the newly made bead to prevent slag inclusion in the weld when the bead is restarted. Restart the arc about 1/4"–3/8" (6 mm–9 mm) ahead of the crater, **Figure 13-4.** Using a slightly longer-than-normal arc, move the arc back to the crater. Hold it until a weld pool of the proper size forms and just fuses into the last ripple. Then lower the

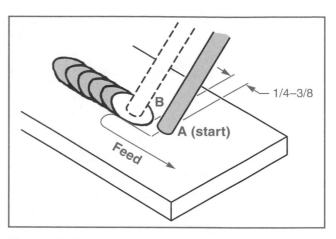

Figure 13-4.
If necessary, restart the arc about 1/4"–3/8" (6 mm–9 mm) ahead of the crater. Using a slightly longer-than-normal arc, move the arc back to the crater and hold it until a weld pool of the proper size forms and just fuses into the last ripple. Lower the electrode to maintain the proper arc length and continue with the bead.

electrode, maintain the proper arc length, and continue with the bead.

Finish the exercise by running continuous beads on all the lines marked on the steel plate. A good job of continuous bead welding is shown in **Figure 13-5.**

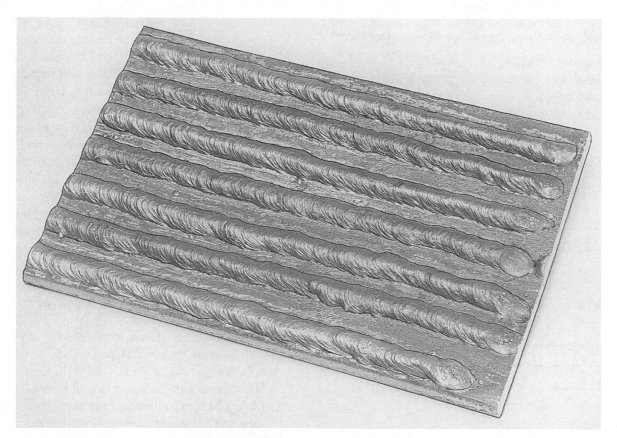

Figure 13-5.
A completed continuous bead exercise. The beads show good penetration and shape, as well as uniform size and appearance.

Name _____ Score _____

Check Your Progress

1. Do not discard used electrodes until they are _____–_____ inches long.

2. Materials used to draw lines on steel plate to indicate where to lay the beads are:

   a. _____

   b. _____

3. List the two reasons you may have to restart the arc.

   a. _____

   b. _____

4. At the end of the weld, you can stop the bead by quickly backing away or _____ the electrode to break the arc.

5. If you have to stop the bead for any reason, what should you do to prevent slag inclusion in the weld when the bead is restarted?

   _____

   _____

   _____

Things to Do

Examine the weld beads you have made on the practice piece. Note the appearance of each bead. Did you encounter problems starting, stopping, or restarting the weld? You may also use **Figure 11-6** as a guide. Write your observations in the appropriate column.

Appearance	Cause	Correction
Example: Very narrow bead	Feed too rapid or amperage setting too low	Slow feed or increase amperage setting

Notes

Unit 14
Running Multidirectional Beads

After completing this unit, you will be able to:
○ Run individual beads from left to right, right to left, top to bottom, and bottom to top.
○ Run a continuous, multidirectional bead that is uniform in shape and size.
○ Identify problems that occur when electrodes that have absorbed moisture are used.

The expert welder can run weld beads in any direction, **Figure 14-1.** To develop this skill, you should practice running a bead from left to right, bottom to top, right to left, and top to bottom.

Study the plans shown in **Figure 14-2** carefully before starting to work. Use 1/8″ (3 mm) diameter E6012, E6013, or E7018 electrodes.

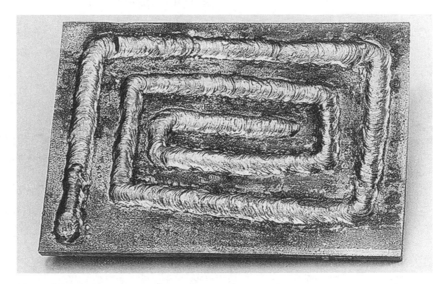

Figure 14-1.
Example of a completed multidirectional welding exercise.

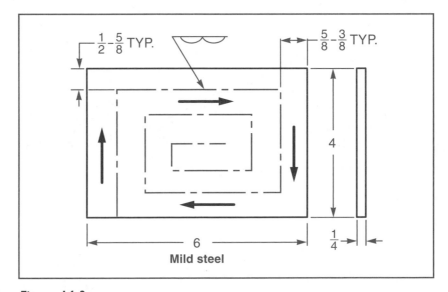

Figure 14-2.
Plan for an exercise in running multidirectional beads.

Use Dry Electrodes

Problems are created when electrodes that have absorbed moisture are used. These electrodes will produce welds that are porous and rough in appearance. If electrodes are allowed to absorb moisture, bake them until they are dry or discard them. A good rule to follow is to take only enough electrodes from storage to last for about an hour.

Procedure:

1. Put on your safety glasses and cut or shear your metal to size. Use mild steel. Carefully remove all burrs and sharp edges.
2. Use soapstone or chalk to draw the series of lines on the steel as specified in Figure 14-2.

3. Check over your safety equipment and dress to begin welding. Turn the welding machine *on* and make the proper settings for polarity and amperage.
4. Position the work and attach the ground clamp.
5. Place the electrode in the holder, strike the arc, and run the bead along the lines as indicated by the arrows.
6. Remove the slag and examine your weld. Compare it with the weld characteristics shown in **Figure 11-7.**

 Warning: Be sure to wear your safety glasses or goggles when removing slag.

Your finished job should look like the example shown in Figure 14-1.

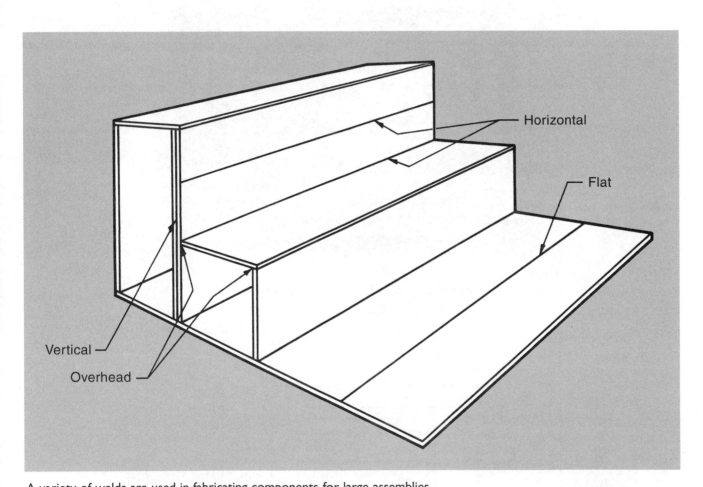

A variety of welds are used in fabricating components for large assemblies.

Name _____ Score _____

Check Your Progress

1. What problems can occur if electrodes are allowed to absorb moisture?

   _____

   _____

   _____

2. List any difficulty encountered when you changed direction of feed.

   _____

   _____

   _____

3. Describe the weld you have made. Include such items as appearance; straightness; and uniformity of the bead, spatter, undercutting, etc.

   _____

   _____

   _____

Things to Do

1. Prepare a list of problems that are created when an electrode that has absorbed moisture is used to make a weld. Secure samples of these problems.

   _____

   _____

   _____

2. Make a sketch for an advanced multidirectional welding exercise. Curves and straight lines may be combined to provide an opportunity for more practice in laying accurate beads.

3. If time permits, perform the actual welding exercise you planned in the previous step.

Notes

Unit 15
Padding

After completing this unit, you will be able to:
○ Explain the circumstances when padding is the most effective means of repair.
○ Run a parallel layer of continuous beads that are relatively smooth, fused to one another, and have only minor depressions between them.
○ Create a second layer of beads, on a right angle to the first layer, using a weaving motion to help maintain a uniform width.

In some situations, it is more economical to repair parts than to replace them. Many parts can be repaired by building up worn surfaces with weld material. Each built-up surface is then machined to required size. This process is called *padding,* or *surfacing,* and it is done by depositing several layers of beads (usually at right angles to each other) until the necessary thickness is attained, **Figure 15-1.** Padding is often used for hard-surfacing, which is covered in Unit 30.

Padding is also very important from the skill development standpoint. Welding over uneven surfaces requires adjustments in forward travel speed when running the second and third passes. You must slow down over depressions and speed up over the high points.

Study the plans in **Figure 15-2,** and cut or shear a mild steel plate to size. Use 1/8″ or 5/32″ (3 mm to 5 mm) dia. E6012 or E6013 electrodes for this exercise.

With this exercise, you will be introduced to the welding technique of *weaving,* which is used to increase the width and volume of the bead. This technique will prove helpful when you pad, or build up, a job. It is also useful when you must make fillet and deep groove welds in later welding exercises.

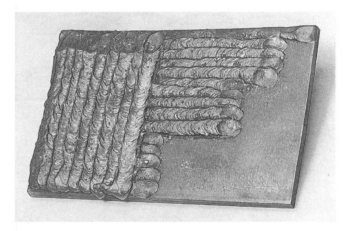

Figure 15-1.
Starting to build up a practice piece as described in this unit.

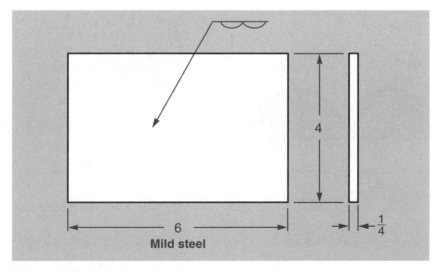

Figure 15-2.
Typical plan for a padding job.

Procedure:

1. Make the usual preparations to weld. Be sure to wear safety glasses.
2. Strike an arc and run the first bead close to and parallel with the long edge of the work. Clean the slag from the weld and run another bead slightly overlapping the first, **Figure 15-3.** Angle the electrode slightly toward the previous bead. Continue the procedure until the work surface is covered. Clean the beads with a wire brush and carefully examine the welded surface. It should have a relatively smooth finish, with the beads fused into one another so there are only minor depressions between them, **Figure 15-4.**

3. Apply the second layer at a right angle to the first. Regulate welding speed for slower travel over depressions and faster movement over the high points. To acquire additional welding skill, use a weaving motion to make the beads on the second layer. Several weaving patterns are shown in **Figure 15-5.** Practice each pattern until you can maintain a bead of uniform width.
4. Build up the surface until it is about 3/4″ (19 mm) thick, square with the base metal, and uniform in thickness.
5. To check the quality of your work, saw through the pad. For easier sawing, let the pad cool slowly in dry sand or ashes. Do not quench it in cold water. Examine the cut. Each layer should be fused with no slag inclusions or holes between the beads.
6. Practice padding until you can produce a solidly built-up piece that is square with the base metal and uniformly thick.

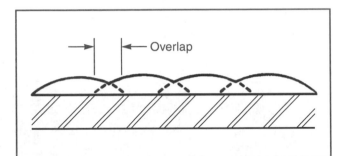

Figure 15-3.
Beads should slightly overlap.

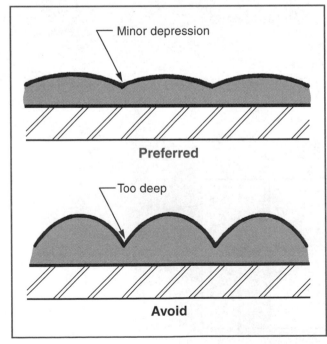

Figure 15-4.
The welds should have a relatively smooth finish with only minor valleys, or depressions, between beads.

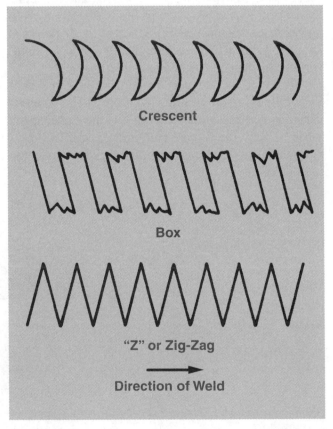

Figure 15-5.
Welding with a weaving motion will make a wider bead.

Name _____ Score _____

Check Your Progress

1. Padding is often used to _____ worn surfaces on machines and equipment.

2. When welding over uneven surfaces, what are two things you should try to do?

   a. _____

   b. _____

3. A technique used to increase the width and volume of the bead is called _____.

4. When padding, the successive layers of beads are usually deposited at _____ to one another.

5. The quality of a completed pad may be determined by _____.

Things to Do

1. Use the space provided below to describe the quality of the weld buildup at the point where the piece was cut. If time permits, make additional cuts and examine these sections.

   _____

   _____

   _____

   _____

   _____

   _____

2. Did the quality of your welds improve as you practiced? What improvements were noted?

   _____

   _____

   _____

3. Describe the appearance of the built-up piece you made. Were the sides square? Was the piece uniform in thickness? Was the final built-up layer the same size as the base metal?

   _____

   _____

   _____

   _____

4. Sketch the following weave patterns:

   Crescent

   Box

   "Z"

Notes

Unit 16
Welding Problems—How to Solve Them

After completing this unit, you will be able to:
- ◯ Identify weld problems.
- ◯ Describe the possible causes of weld problems.
- ◯ List the appropriate solutions for various weld problems.

By this time, you have probably encountered several of the most common welding problems. These problems, their causes, and solutions are presented in this unit so they will be handy for easy reference.

Poor Appearance	Causes	How to Solve
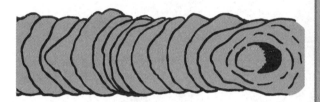	1. Current setting too high or too low.	1. Correct current setting.
	2. Wrong type of electrode.	2. Use proper electrode.
	3. Faulty electrode.	3. Check electrodes before use.
	4. Overheated work.	4. Allow work to cool between passes.
	5. Incorrect speed of travel.	5. Adjust speed so that proper bead Is formed.
	6. Electrode manipulated improperly.	6. Use proper welding technique.

Excessive Spatter	Causes	How to Solve
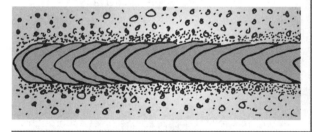	1. Current setting too high.	1. Use correct current setting.
	2. Arc too long.	2. Adjust to proper length arc.
	3. Arc blow.	3. Minimize arc blow.
	4. Wrong polarity for electrode being used.	4. Use correct electrode and polarity.
	5. Faulty electrode.	5. Select suitable electrode.

Arc Hard to Start	Causes	How to Solve
	1. Current setting too low.	1. Correct current setting.
	2. Work not cleaned.	2. Clean work.
	3. Work not properly grounded.	3. Clamp ground solidly to bare metal.
	4. Flux covered electrode tip.	4. Clean electrode tip.

Poor Fusion	Causes	How to Solve
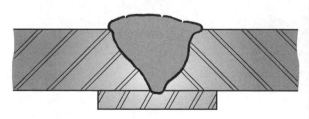	1. Current setting too low.	1. Correct welding current.
	2. Incorrect welding speed.	2. Adjust welding speed to ensure melting of both sides of joint.
	3. Wrong type electrode.	3. Use proper electrode.
	4. Arc too long.	4. Hold correct length arc.
	5. Work not properly prepared for welding.	5. Make sure joint is clean. "V" or groove joint if necessary.

Undercutting

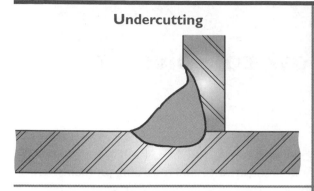
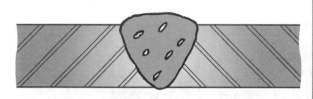

Causes
1. Current setting too high.
2. Welding speed too fast.
3. Arc too long.
4. Wrong size electrode.
5. Incorrect electrode to work angle.

6. Faulty electrode manipulation.

How to Solve
1. Correct welding current.
2. Reduce speed of travel.
3. Hold correct length arc.
4. Use correct size electrode.
5. Adjust electrode angle so that arc force will hold molten metal until undercut fills.
6. Use a uniform weave.

Porous Welds

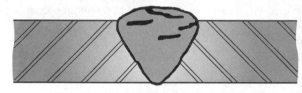

Causes
1. Short arc.
2. Welding speed too fast.
3. Welding speed too slow.
4. Insufficient time in molten state.
5. Impurities in or on base metal.
6. Wrong type electrode.

How to Solve
1. Hold correct length arc.
2. Reduce speed of travel.
3. Increase speed of travel.
4. Allow enough time for gases to escape.
5. Clean base metal thoroughly.
6. Use proper electrode.

Slag Inclusion

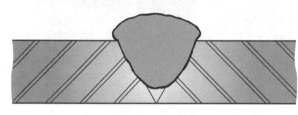

Causes
1. Current setting too low.
2. Arc too short.
3. Welding speed too slow.
4. Welding speed too fast.
5. Faulty electrode manipulation.

How to Solve
1. Correct welding temperature.
2. Hold correct arc length.
3. Increase welding speed.
4. Decrease welding speed.
5. Use correct electrode-to-work angle so that arc force prevents molten metal from overtaking slag.

Incomplete Penetration

Causes
1. Welding speed too fast.
2. Electrode too large.
3. Current setting too low.

4. Impurities in or on base metal.
5. Weld groove not proper size.

How to Solve
1. Weld more slowly.
2. Select electrode according to welding groove size.
3. Correct welding current.
4. Clean base metal thoroughly.
5. Allow sufficient space at bottom of joint.

Cracked Welds

Causes
1. Wrong type electrode.
2. Base metal high carbon steel.
3. Weld cooled too rapidly.
4. Work too rigid.

5. Weld and part sizes unbalanced.

How to Solve
1. Use proper electrode.
2. Cool work slowly.
3. Cool work slowly.
4. Design work to eliminate rigid joints.
5. Heat parts before welding. Cool slowly after welding.

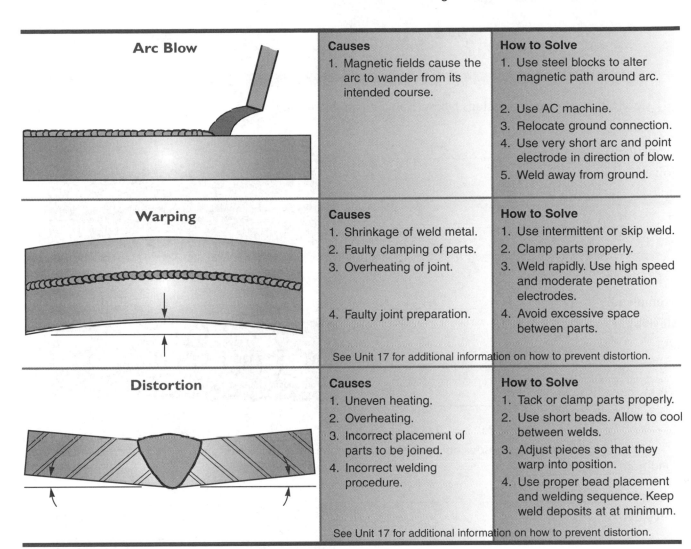

Arc Blow	Causes	How to Solve
	1. Magnetic fields cause the arc to wander from its intended course.	1. Use steel blocks to alter magnetic path around arc. 2. Use AC machine. 3. Relocate ground connection. 4. Use very short arc and point electrode in direction of blow. 5. Weld away from ground.
Warping	**Causes**	**How to Solve**
	1. Shrinkage of weld metal. 2. Faulty clamping of parts. 3. Overheating of joint. 4. Faulty joint preparation.	1. Use intermittent or skip weld. 2. Clamp parts properly. 3. Weld rapidly. Use high speed and moderate penetration electrodes. 4. Avoid excessive space between parts.
	See Unit 17 for additional information on how to prevent distortion.	
Distortion	**Causes**	**How to Solve**
	1. Uneven heating. 2. Overheating. 3. Incorrect placement of parts to be joined. 4. Incorrect welding procedure.	1. Tack or clamp parts properly. 2. Use short beads. Allow to cool between welds. 3. Adjust pieces so that they warp into position. 4. Use proper bead placement and welding sequence. Keep weld deposits at at minimum.
	See Unit 17 for additional information on how to prevent distortion.	

Name _____ Score _____

Check Your Progress

Carefully study the illustrations below. Name the problem and list two or three ways to correct the welding condition.

Example of: _____

Solve by:

1. _____.

2. _____.

3. _____.

Example of: _____

Solve by:

1. _____.

2. _____.

3. _____.

Carefully study the illustrations below. Name the problem and list two or three ways to correct the welding condition.

Example of: _____

Solve by:

1. _____.

2. _____.

3. _____.

Example of: _____

Solve by:

1. _____.

2. _____.

3. _____.

Example of: _____

Solve by:

1. _____.

2. _____.

3. _____.

Things to Do

1. Look around the shop or your home for items or products that have been arc welded. Carefully examine each weld to see if any welding problems exist. Make a list of the items and describe any defects found. Suggest ways in which each welding problem could have been corrected.

    _____

    _____

    _____

    _____

    _____

    _____

2. Select a piece of scrap steel and deposit a bead at least 3″ (8 cm) long. Use a very high current setting for the electrode being used. Chip away the slag and observe the bead. List all the problems caused by the excessive current setting.

    _____

    _____

    _____

    _____

3. Perform the same experiment as in number 2 above, except deposit the bead at a very slow rate of speed. Chip away the slag and note the welding problems you observe.

    _____

    _____

    _____

    _____

Notes

Unit 17
Controlling Distortion

After completing this unit, you will be able to:
○ List the four general types of distortion.
○ Identify the types of distortion in weld samples.
○ Explain the preventive actions for each type of distortion.

When you welded your practice pieces, you probably noted that the metal often became distorted (twisted or warped out of shape). This distortion is caused by the metal expanding (getting larger) as it is heated and contracting (getting smaller) as it cools, **Figure 17-1.**

Under normal conditions, heated metal will return to its original shape and size when it cools. However, metal heated by the welding operation cannot return to its original size because the weld bead acts as a restraint, holding the metal in the distorted position.

While expansion and contraction cannot be prevented, they can be controlled. Distortion is affected by many factors. You must recognize when distortion will occur and understand how to minimize it. The quality of a welded joint is determined by how well the welder controls expansion and contraction of the metal as it is welded.

Types of Distortion

There are four general types of distortion that may affect the welded section or joint. By understanding the forces that cause distortion, you will be able to use these forces to your advantage to pull the welded section or joint into the required position.

Angular distortion is illustrated in **Figure 17-2.** Contraction of the weld as it cools pulls the plates out of alignment and accounts for distortion of this type.

Angular distortion can be held to a minimum using one of the following techniques:
1. Position the pieces slightly out of alignment so the shrinking weld deposit will pull the members into position, **Figure 17-3.**

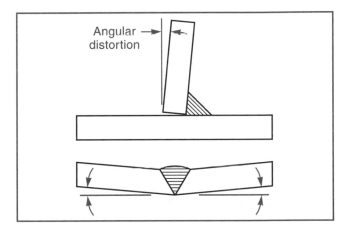

Figure 17-2.
In angular distortion, the weld bead contracts as it cools and pulls the parts being welded out of alignment.

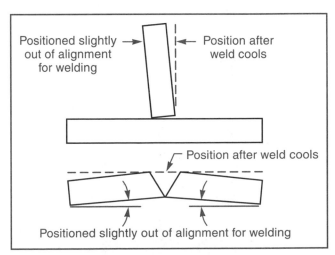

Figure 17-3.
Angular distortion can often be controlled by positioning the pieces slightly out of alignment so the shrinking weld pulls them into position.

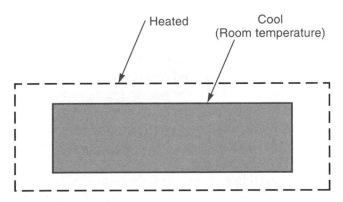

Figure 17-1.
Almost all metals expand when they are heated and contract when they are cooled.

2. Use as little weld material as possible. Do not overweld, since a large bead will increase contraction.
3. Develop a welding sequence that builds up the weld equally on each side of the joint, **Figure 17-4.**

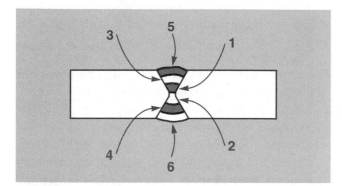

Figure 17-4.
Angular distortion can also be controlled by developing a welding sequence that builds up the weld equally on each side of the joint.

4. Utilize staggered beads as shown in **Figure 17-5** and allow the contracting forces to balance each other.

Longitudinal (lengthwise) distortion, **Figure 17-6,** causes the metal to bend upwards in the direction of the weld. To correct this distortion problem:

1. Use "backstep" welding and intermittent welding, **Figure 17-7.**
2. Utilize small diameter electrodes and low amperage.

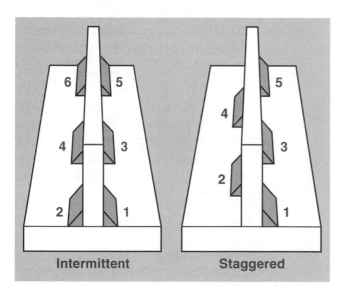

Figure 17-5.
Intermittent and alternating beads can also be utilized to control distortion.

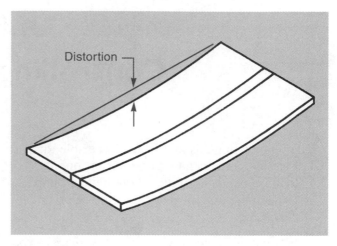

Figure 17-6.
Longitudinal distortion causes the metal to bend upwards in the direction of the weld.

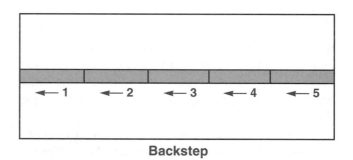

Backstep

Intermittent

Figure 17-7.
Backstep and intermittent welding can be used to control longitudinal distortion.

3. Peen the weld bead with a hammer. This stretches the bead and counteracts its tendency to shrink as it cools. Do not overpeen, as this may cause the weld to crack.
4. Mount the parts to be welded in a fixture. This will hold them rigid as they cool. Since there is a tendency for some welded parts to contract when removed from the fixture, the fixture is designed to position the parts slightly out of alignment. When the parts are removed from this type of fixture, the contraction forces pull the parts into exact alignment.

Traverse (crosswise) distortion occurs when the cooling weld causes butt-welded pieces to draw together and sometimes overlap, **Figure 17-8.**

Traverse distortion can be controlled or minimized in several ways. One method involves tack welding the work pieces in position before running the full weld, **Figure 17-9.** Tack weld spacing should be about 4″ on center for thin materials (1/8″ or less). Thicker material provides more rigidity, so fewer tack welds are needed. Another method used to control traverse distortion is spacing the plates as shown in **Figure 17-10.** If the plates are fairly thick, placing a wedge in the joint will maintain the proper spacing, **Figure 17-11.**

Bowing, or *Buckling,* **Figure 17-12,** often occurs when sheet metal or thin sections are welded. These problems are caused by the uneven expansion and contraction of the welded area.

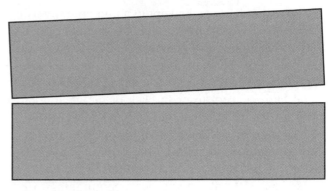

Figure 17-10.
Spacing the plates can also control or minimize traverse distortion. The plates are spread about 1/8″ for each foot of length of the pieces being welded.

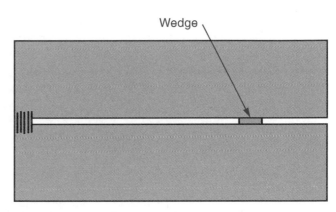

Figure 17-11.
A wedge placed near the end of the joint will also help control traverse distortion.

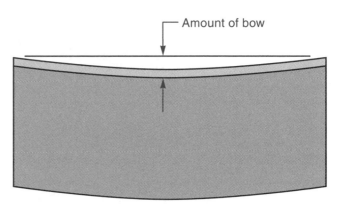

Figure 17-12.
This piece has bowed.

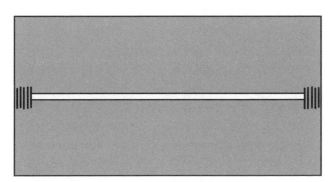

Figure 17-8.
Traverse distortion occurs when the cooling weld causes butt-welded pieces to draw together and sometimes overlap.

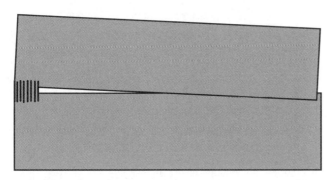

Figure 17-9.
You can control or minimize traverse distortion by tack welding the plates before running the full weld.

The problem can be kept to a minimum by using electrodes designed for welding sheet metal, such as E6013. Maintaining a short arc and/or using a chill bar can also help control bowing and buckling. A *chill bar,* **Figure 17-13,** is a bar of copper or other metal laid adjacent to or in back of the weld area to carry off the excess heat that causes the distortion.

Distortion can cause many problems. However, through careful planning, the welder can control distortion and keep it to a minimum. In the next few units, you will have numerous opportunities to put your knowledge and skill to use to minimize the problems of distortion.

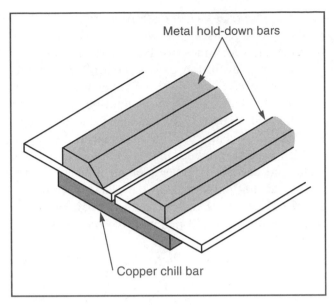

Figure 17-13.
Chill bars and metal hold-down bars will help control bowing, or buckling, when thin sections are welded.

Name _____ Score _____

Check Your Progress

1. When metal is heated, it _____.

2. Metal _____ as it cools.

3. Distortion is caused by:

   _____

   _____

4. The quality of a welded joint is determined by:

   _____

   _____

   _____

5. Which of the four general types of distortion is illustrated below? _____ Using the partial drawing to the right, sketch in one way to control this problem.

6. Which type of distortion is illustrated below?

   _____

7. The above distortion problem can be kept to a minimum by: Check the correct answer(s).

   a. _____ using small diameter electrodes and low amperage.

   b. _____ peening the weld bead with a hammer.

   c. _____ gas welding.

   d. _____ All of the above.

   e. _____ None of the above.

8. Sketch in what is meant by the term *intermittent* welding.

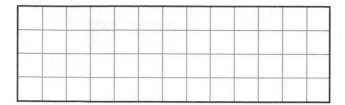

9. Sketch in what is meant by the term *backstep* welding.

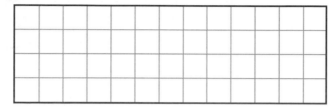

10. Which type of distortion is illustrated below?

_____

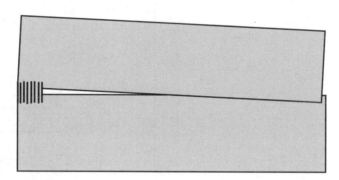

11. The above distortion problem can be controlled by:

    Check correct answer(s).

    a. _____ tack welding.

    b. _____ careful spacing of the plates.

    c. _____ using a wedge to maintain proper spacing.

    d. _____ All of the above.

    e. _____ None of the above.

12. Sketch in what is meant by the term *tack weld.*

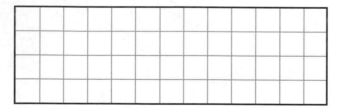

13. _____ or _____ is the fourth type of distortion.

14. A *chill bar* is a bar of copper or other metal designed to:

    _____

    _____

Name _____ Score _____

Things to Do

1. Prepare models that illustrate distortion in welding. Use heavy cardboard for metal parts and glue the pieces together. Soft clay can be rolled into thin rods and set in place to show the welds. The final models can be painted gray to make them look like metal. Choose as many distortion problems as you have time to prepare.

2. To further understand expansion and contraction of welded pieces, select a scrap piece of steel and deposit a bead at least 4″ (10 cm) long. Immediately dip the piece in cold water. After cooling, carefully examine the piece for distortion and record your findings.

3. Visit a local welder or contractor. Ask about the types of distortion problems encountered during regular welding activities. Also, ask how these problems are overcome. Report your discussion to the class.

Notes

Unit 18
Welding Square-Groove Joints

After completing this unit, you will be able to:

○ Identify four joint types that may be used in conjunction with square-groove joints.

○ Complete a single-pass, square-groove, closed butt joint with acceptable weld integrity and penetration.

○ Produce a single-pass, square-groove, open butt joint with uniform appearance and acceptable weld integrity.

The square-groove weld has many applications. It can be used with butt (closed and open), tee, corner, and edge joints, **Figure 18-1.** The exercises described in this unit will be concerned with *square-groove butt joints.*

The butt joint is fairly easy to prepare. The plates need only be matched along the square edges of the metal being welded, **Figure 18-2.**

If the plates do not exceed 3/16″ (5 mm) in thickness, the joints can be welded in a single pass. The weld is usually made from the side that shows from the finished side. However, complete penetration by welding from both sides can be made on metal up to 1/8″ (3 mm) thick without a root opening (closed butt joint). When welding from both sides on materials up to 1/4″ (6 mm) thick, place the edges slightly apart

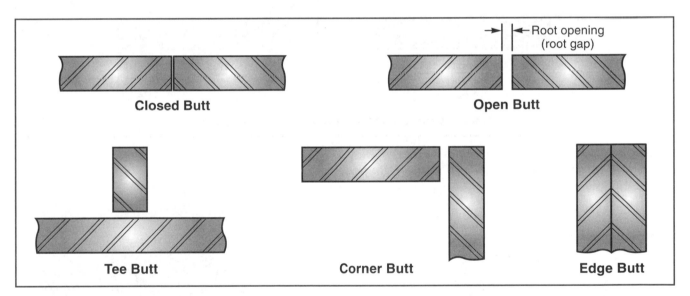

Figure 18-1.
Square-groove joints.

Figure 18-2.
The plates should be in good alignment along the edges that are to be welded.

(open butt joint) for complete penetration. Spacing of the plates depends upon the material thickness, **Figure 18-3.**

Single-Pass, Closed Butt Joint Groove Weld

The edges of the *closed butt joint groove weld* are placed together. Refer to Figure 18-3 to help you determine the spacing.

Study the plans carefully and cut or shear the metal to size, **Figure 18-4.** Remove all burrs and sharp edges. Use 1/8″ (3 mm) diameter E6010 or E6011 electrode. The completed exercise is shown in **Figure 18-5** for comparison with your work.

Procedure:

1. Make the usual preparations to weld. Be sure to wear safety glasses.

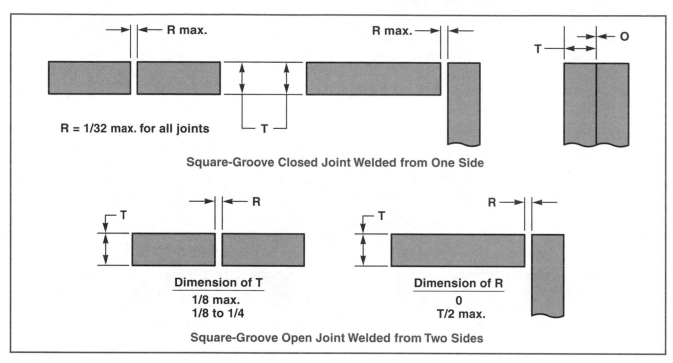

R = 1/32 max. for all joints

Square-Groove Closed Joint Welded from One Side

Dimension of T
1/8 max.
1/8 to 1/4

Dimension of R
0
T/2 max.

Square-Groove Open Joint Welded from Two Sides

Figure 18-3.
Information for determining the spacing for square-groove joints.

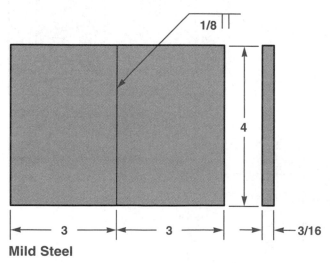

Mild Steel

Figure 18-4.
Plans for a square-groove, closed butt joint.

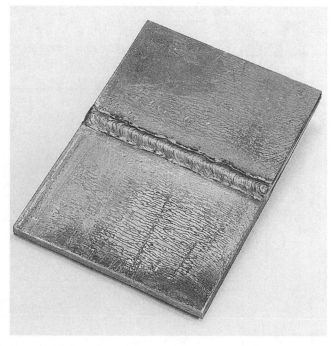

Figure 18-5.
Completed single-pass, square-groove, closed butt joint.

2. Place the two plates in position on the worktable with two edges in close contact.

3. Tack weld the plates together. A tack weld is a short weld made near the ends of the work to hold the plates together while they are welded, **Figure 18-6**.

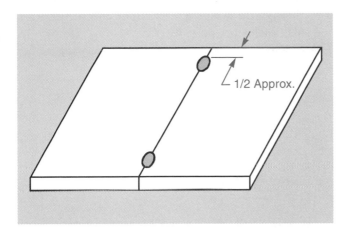

Figure 18-6.
Tack welds are made approximately 1/2″ (13 mm) in from each end.

4. The single-pass weld is made in much the same manner as the continuous bead weld. Use care to keep the electrode centered on the joint. In this way, half the weld will be deposited on each plate. To eliminate the crater at the end of the weld, hold the electrode stationary long enough to fill the crater and then withdraw it gradually.

5. Allow the weld to cool and remove the slag.

6. To check the depth of penetration and integrity of the weld, cut the test piece approximately 2″ wide across the weld. The width of the test piece should not exceed 1 1/2″. Put one piece in the vise as shown in **Figure 18-7**. Using a large hammer, bend the test piece over until the weld is broken. Examine the cut section and broken section for imperfections such as porosity, incomplete fusion, lack of penetration, slag inclusion, undercut, etc.

7. Improve your skill by making additional closed butt joint welds. Allow the welds to cool, remove the slag, cut the pieces in half across the welds, and break the joints. Examine the welds for imperfections.

8. When you are able to make acceptable closed butt joint welds, make several more but this time experiment. Make one joint:
   - using a weaving motion when running the bead.
   - using increased amperage.

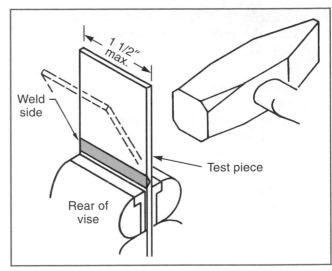

Figure 18-7.
Use an old vise to hold the work when you bend it over to break the joint. The welded piece may have to be moved to the anvil to break the joint.

   - using decreased amperage.
   - using E6012 or E6013 electrodes.

Break the joints and examine the welds. How do they compare with the other practice pieces?

Single-Pass, Open Butt Joint Weld

The edges of the *open butt joint weld* are placed slightly apart. See **Figure 18-3** to help you determine the spacing.

Study the plans in **Figure 18-8** carefully, and cut or shear the metal to size. Remove all burrs and sharp edges. Use 1/8″ (3 mm) diameter E6010 or E6011 electrodes.

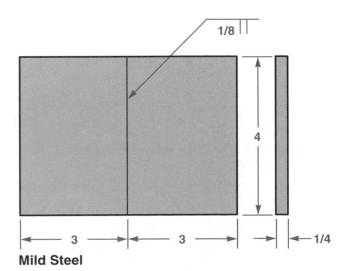

Mild Steel

Figure 18-8.
Plans for a square-groove, open butt joint.

Procedure:

1. Make the usual preparations to weld. Be sure to wear safety glasses.
2. Position the plates, spacing them properly, and tack weld them together.
3. Run the weld, being sure to deposit material equally on both plates. A slight weaving motion is recommended to handle the extra width of the opening.
4. Allow the piece to cool, clean the weld, and cut an approximately 2″ wide test piece across the weld.
5. Inspect the weld for uniform appearance, checking the other side of the plate for penetration.
6. Break the test piece and check for weld integrity.
7. Make additional practice pieces but vary the opening between the plates.
8. When you can make acceptable square-groove open butt joint welds, weld several additional pieces from both sides. Cut the pieces lengthwise through the welds and examine the sections for slag inclusion, porosity, etc.

Name _____ Score _____

Check Your Progress

1. List four applications of the square-groove weld joint.

   a. _____

   b. _____

   c. _____

   d. _____

2. If the plates do not exceed _____ ″ in thickness, the joints can be welded in a single pass.

3. On an open butt joint, the spacing of the plates depends upon their _____ .

4. To eliminate the _____ at the end of the weld, hold the electrode stationary for a moment and then withdraw it gradually.

5. The single-pass weld is made in much the same manner as the _____ weld.

6. Examine the cut section and broken section of a weld for imperfections such as:

   a. _____

   b. _____

   c. _____

   d. _____

   e. _____

Things to Do

1. The American Welding Society has established *Standards of Acceptability* for weld soundness. The standards for the welds just made are:

   - *Contour*—The exposed face of the weld shall be reasonably smooth and regular. There shall be no overlapping or undercutting.

   - *Extent of fusion*—There shall be complete fusion between the weld and base metal and full penetration to the root of the weld.

   - *Soundness*—The weld shall contain no gas pocket, oxide particle, or slag inclusion exceeding 1/8″ in its greatest dimension. In addition, no square inch of weld metal area shall contain more than six gas pockets exceeding 1/16″ in their greatest dimension.

   Prepare a chart using the above standards and grade your welds according to them. Number the test pieces in the order they were made.

2. Secure samples of butt welds made by a professional welder.

3. Contact a local company that does extensive welding. Secure a copy of tests they require of their welders before they are hired.

4. Contact a local company that does extensive welding and secure a copy of the welding standards they use.

Notes

Unit 19
Single-Fillet Lap-Joint Welds

After completing this unit, you will be able to:
- ○ Create a single-fillet lap-joint weld that is consistent.
- ○ Explain the factors used to determine the length and strength of a fillet weld.
- ○ Properly utilize a fillet gauge to check the weld size.

The lap joint, **Figure 19-1,** like the butt joint, requires little prewelding preparation. The plates need only be reasonably clean, flat, and in close contact with each other. The weld used to join a lap joint is called a *fillet* (fil'-it) *weld,* **Figure 19-2.**

Study the plans in **Figure 19-3** carefully. Cut or shear the metal to size. Remove all burrs and sharp edges. Check plates for flatness. Use 1/8″ diameter E6012 or E7018 electrodes. The completed exercise is shown in **Figure 19-4.**

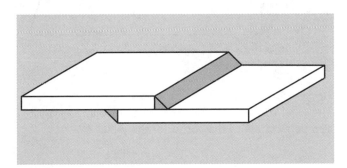

Figure 19-1.
Typical lap joint. Note that the joint has been welded in two places.

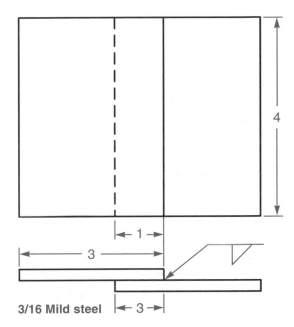

Figure 19-3.
Plans for single-fillet lap-joint exercise.

3/16 Mild steel

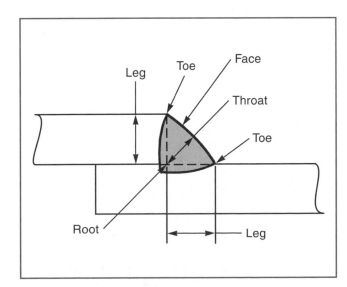

Figure 19-2.
Fillet weld.

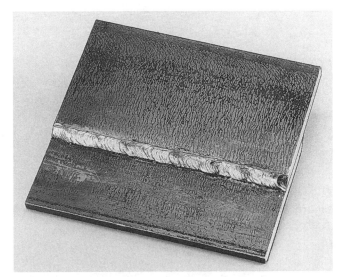

Figure 19-4.
Completed single-fillet lap-joint weld.

Procedure:

1. Make the usual preparations to weld. Be sure to wear safety glasses.
2. Position the plates on the worktable and tack weld them together.
3. Run the weld by holding the electrode at about a 45° angle, **Figure 19-5.** The arc should be directed into the corner of the joint.
4. Carefully observe the bead as you run it. Be sure it penetrates to the root of the joint and builds up to the top of the lapping plate. Direct the arc more to the bottom plate if you start cutting back the edge of the top plate. See **Figure 19-6** for a properly made fillet weld.

5. Cool the plate and remove the slag.
6. Cut the plate in half across the weld. Examine the weld for uniformity and smoothness. The weld should penetrate both plates evenly and also penetrate into the corner of the joint, **Figure 19-7.**
7. A more thorough inspection of the penetration can be made by breaking the weld, **Figure 19-8.**

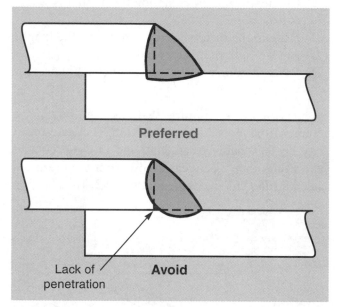

Figure 19-7.
The weld should penetrate both plates evenly and must also penetrate into the corner of the joint.

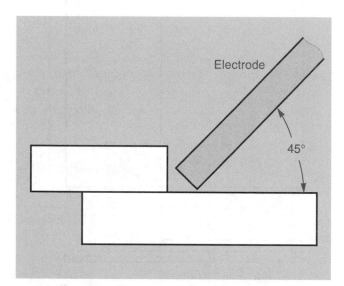

Figure 19-5.
Holding the electrode at a 45° angle when running a single-pass fillet weld.

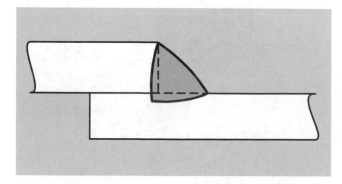

Figure 19-6.
A properly made fillet weld.

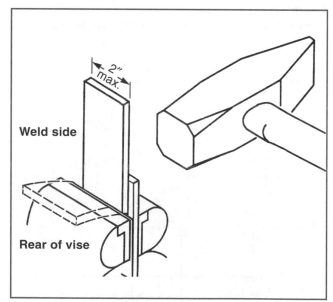

Figure 19-8.
Inspect the quality of the weld by bending the top plate over until it breaks or is flattened.

8. Make additional practice pieces until you can produce acceptable fillet welds. Weld both sides of the practice pieces. Cut them across the welds at several points to check that the weld penetrates both plates evenly and also penetrates into the corner of the joint.

Note: The strength of a fillet weld is determined by the thickness of the weld ***throat***. The weld size is governed by the lengths of the ***legs***. The actual weld can be up to 25% larger than specified and still meet specifications. However, it cannot be smaller than the dimensions specified. A flat or slightly convex face is preferred, **Figure 19-9**.

Fillet weld sizes can be easily checked with a ***fillet gauge***. The gauge can be shop made, **Figure 19-10**, or purchased from a welding supply house, **Figure 19-11**.

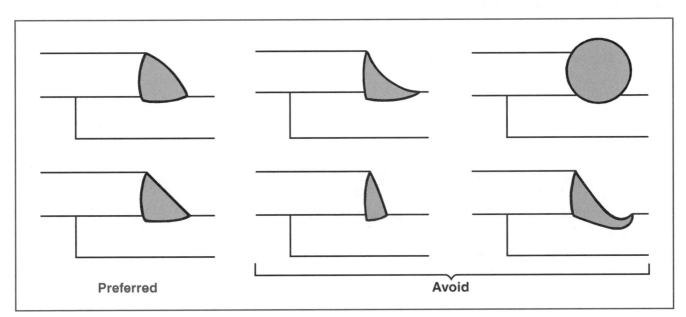

Preferred   **Avoid**

Figure 19-9.
A fillet weld should have a flat or slightly convex face.

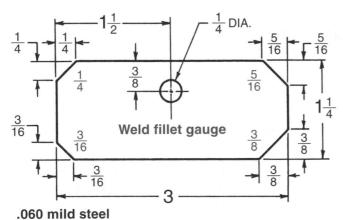

.060 mild steel

Figure 19-10.
A shop-made fillet gauge. Use 1/8″ high steel letters to stamp the figures shown in color. You may want to put your name on the other side of the tool.

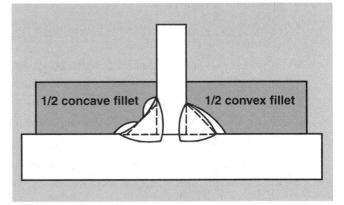

Figure 19-11.
Commercially made fillet gauges.

Name _____ Score _____

Check Your Progress

1. To run a single-fillet lap-joint weld, hold the electrode at about a _____° angle.

2. The arc should be directed into the _____ of the joint.

3. The strength of a fillet weld is determined by the thickness of the weld _____.

4. A fillet weld should have a flat or slightly _____ face.

5. After cutting the plate in half across the weld, examine the weld for:
   a. _____
   b. _____

6. The weld should penetrate both plates:
   a. _____
   b. _____

Things to Do

1. The drawing below shows two plates in position for welding. Draw in the fillet weld and indicate its various parts.

2. A fillet weld soundness test has been developed by AWS. Their *Standard of Acceptability* includes:
   - *Contour*—The exposed face of the weld shall be reasonably smooth and regular. There shall be no overlapping or undercutting. The weld shall conform to the required cross section for the size of weld specified per gauge.
   - *Extent of fusion*—There shall be complete fusion between the weld and the base metal and full penetration to the root of the weld.
   - *Soundness*—The weld shall contain no gas pocket, oxide particle, or slag inclusion exceeding 3/32″ in its greatest dimension. In addition, no square inch of weld metal area shall contain more than 6 gas pockets exceeding 1/16″ in their greatest dimension.
   a. Using the above standards, develop a chart and grade the fillet welds you have made.
   b. List the practice pieces in the order they were made.
   c. Obtain some examples of fillet welds from a local company that does extensive welding. If you are unable to get welds from a local company, look at a product with fillet welds. A piece of construction equipment (backhoe, bulldozer), trailer, or a machine tool will have fillet welds. Compare your welds to the welds made at the local industry or the welds on the equipment.

4. Make a fillet gauge. Use the plan on page 101.

5. Draw the correct welding symbol for the following fillet welds:

Unit 20
Single-Pass Fillet Welds

After completing this unit, you will be able to:

○ Demonstrate the correct electrode positioning for various joint angles.

○ Produce a single-pass fillet weld on a joint that meets the AWS soundness test.

○ List the weld characteristics included in the AWS *Standard of Acceptability* for single-pass fillet welds.

The technique for making a fillet weld is similar to that used for the lap-joint fillet weld. The weld in this exercise is also made in a single pass.

Remember that when making a single-pass weld, the electrode is positioned so it splits the angle of the joint, **Figure 20-1.** Study the plans carefully, **Figure 20-2.** Use 1/8″ or 5/32″ diameter E6012 or E7018 electrodes.

Procedure:

1. Make the usual preparations to weld. Be sure to wear safety glasses.
2. Set up the plates and tack weld to hold them in alignment.
3. Make the weld in much the same way you made the lap-joint weld.

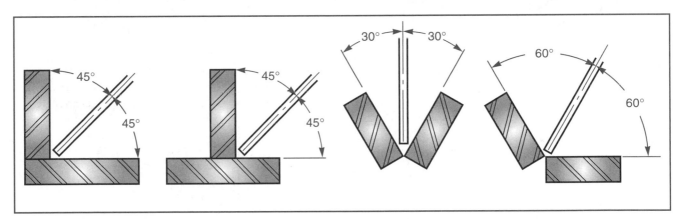

Figure 20-1.
Electrode position for making a single-pass fillet weld. Note that the electrode position splits the angle.

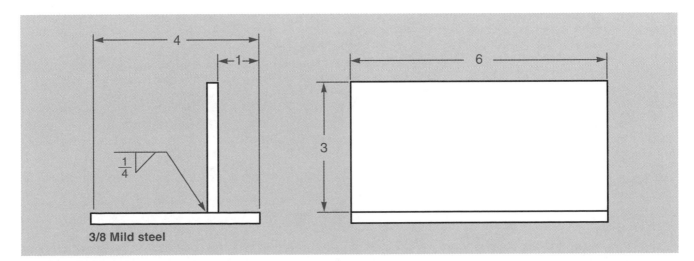

Figure 20-2.
Plans for single-pass fillet weld exercise.

4. Clean and examine the weld, **Figure 20-3.** Check it for size. Use the same *Standard of Acceptability* for weld soundness that was specified for the weld made on the lap joint to evaluate this fillet weld.

5. Examine the plates for distortion.

6. After you can run satisfactory fillet welds, prepare a test joint. Place the welded plates in a press. If a press is not available, put the assembly on an anvil and hit it with a sledge hammer until the weld breaks or the plates are flattened, **Figure 20-4.** If the weld breaks, examine it for weak points and determine what caused them: porosity, poor penetration, etc.

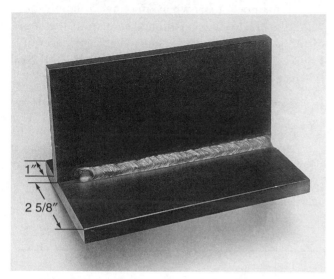

Figure 20-3.
This weld has been cleaned and is ready for examination.

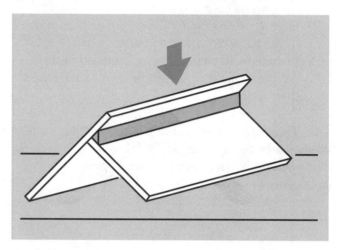

Figure 20-4.
Check the integrity of the weld by applying force as shown until the joint breaks or the plates are flattened. This may be done in a press or with a sledgehammer.

Name _____   Score _____

Check Your Progress

1. The technique used for making a fillet weld is similar to that used for the _____ weld.

2. When making a single-pass weld, the electrode is positioned so that it _____ the angle of the joint.

3. When setting up the plates, why are they tack welded before beginning the single-pass fillet weld?

   _____

4. What size and type (number) electrode should be used when making a single-pass fillet weld?

   _____

Things to Do

1. Sketch in the proper electrode position for making single-pass welds on the following plates.

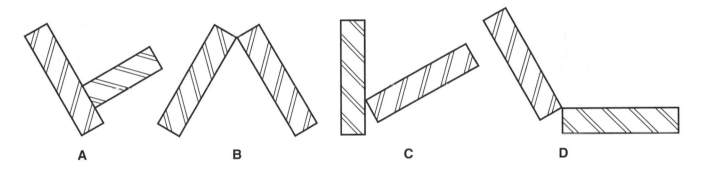

    A   **B**   **C**   **D**

2. Sketch in the welding symbols for the welds shown.

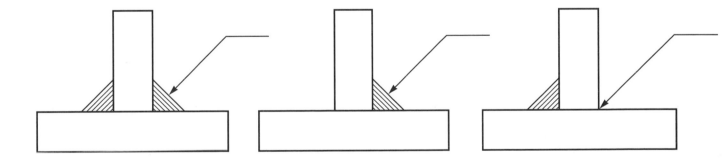

3. Secure samples of properly made fillet welds.

4. In the space below, list the problems you encountered making single-pass fillet welds. Opposite the problems, list techniques that will help you to overcome the problems. See Unit 16.

Problems	Corrections

Unit 21
Multiple-Pass Fillet Welds

After completing this unit, you will be able to:

○ Use stringer beads to make a multiple-pass fillet weld that has equal legs and meets the appropriate integrity standards.

○ Use a weave pattern to make a multiple-pass fillet weld that has equal legs and meets the appropriate integrity standards.

○ Make sound multiple-pass fillet welds with the work tilted at various angles.

Fillet welds 3/8″ and smaller can usually be made with a single pass of the electrode. However, it is more economical to use multiple-passes to make the larger fillet welds required on heavier plate.

Two techniques for making multiple-pass fillet welds are described in this unit. In one method, the fillet weld is built up with beads placed against and on top of one another using a stringer bead technique, **Figure 21-1.** In the other method, multiple passes, plus a weaving motion of the electrode to lay the beads, builds up the fillet, **Figure 21-2.** For maximum strength, fillet welds should be made on both sides of the joint, **Figure 21-3.**

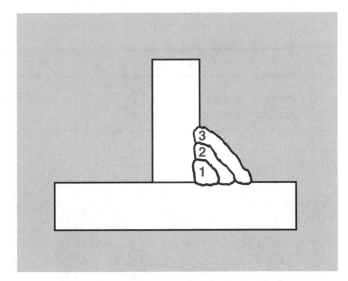

Figure 21-2.
Fillet weld built up of layered welds.

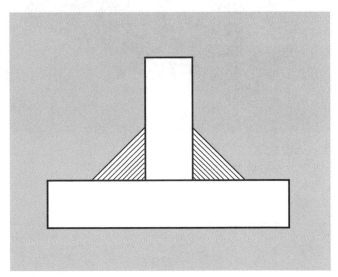

Figure 21-3.
For maximum strength, fillet welds should be made on both sides of the joint.

Figure 21-1.
Fillet weld built up with beads placed against and on top of one another.

Study the plans in **Figure 21-4** and secure the necessary metal. Make multiple-pass fillet welds with E6012, E6013, or E7018 electrodes. Those used on this job should be 1/8″ or 5/32″ (3 mm or 4 mm) in diameter. **Figure 21-5** shows a partially completed weld with which you can compare your own work.

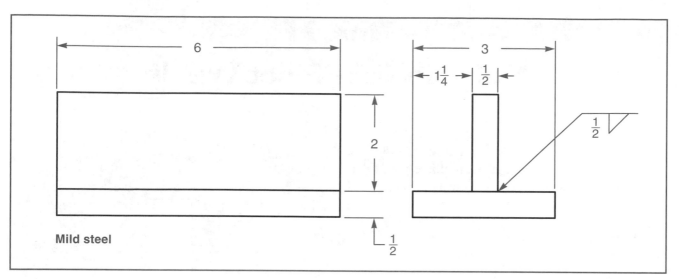

Figure 21-4.
Plans for this job.

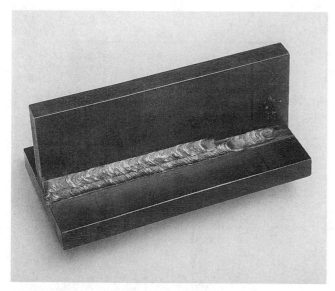

Figure 21-5.
Partially completed multiple-pass fillet weld.

Procedure: *Method 1*

1. Make the usual preparations to weld. Be sure to wear safety glasses.
2. Set up the plates and tack weld them in position.
3. Make the weld. The electrode should be held at a work angle of 45° for the first pass. In this way, equal force from the arc is directed against each plate. The second bead is made on the horizontal plate to form a flat edge on which the third pass can be laid. The third bead is deposited against the vertical plate. See **Figure 21-6** for the different electrode angles needed for the various passes for horizontal fillet welds. In all passes, the electrode is tilted in the direction of travel (approximately 80°). Use care when running the second and third beads so there is no undercutting of the plates, **Figure 21-7.** Any number of layers of beads may be built up in this manner, **Figure 21-8.**

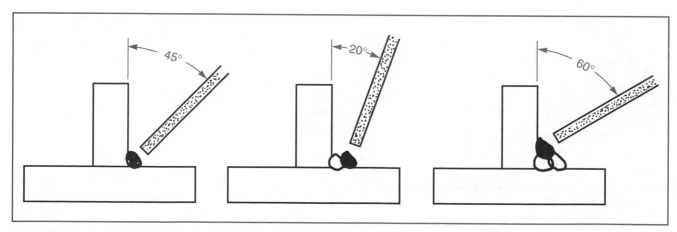

Figure 21-6.
Position of electrode when building up a multiple-pass fillet weld.

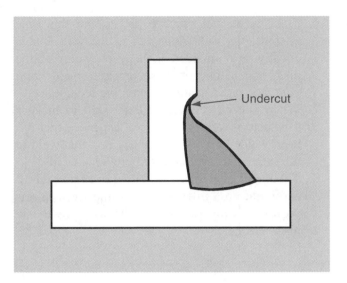

Figure 21-7.
Use care so there will be no undercutting when running the last bead.

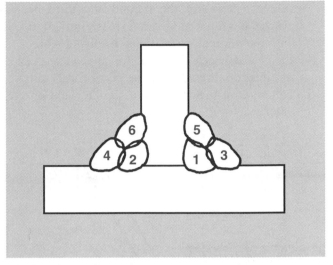

Figure 21-9.
Distortion can be kept to a minimum by using an alternate sequence to build up the weld.

4. Clean the welds and check their integrity. Inspect the plates for distortion.
5. To conserve material, make a second weld *on the other side* of the joint.
6. Make additional practice pieces. However, this time alternate the sequence so the weld builds up equally on each side of the joint, **Figure 21-9.**
7. Compare the distortion of the second practice piece against that of the first piece.
8. When you have developed the skill to make a satisfactory multiple-pass fillet weld, make up a test piece that has been welded on *one* side only. Test the weld, using the AWS standards of acceptability.

9. If test equipment is not available to handle a plate of this thickness, cut a section 1 1/2″ (38 mm) wide from the practice piece, **Figure 21-10.** Hit it with a sledgehammer until it breaks or is flattened.

Procedure: *Method 2*

1. Make the usual preparations to weld. Be sure to wear safety glasses.
2. Set up the plates and tack weld them in position.
3. Make the weld. Use a fairly high current and feed when running the first bead. It is made in the corner of the joint and you need not be concerned if there is some undercutting. The

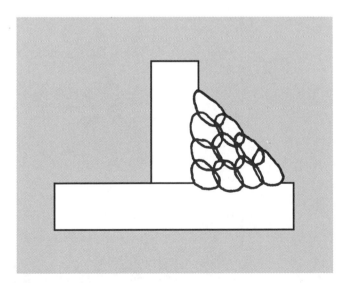

Figure 21-8.
Multiple-layer fillet weld. Almost any size fillet weld can be made in this manner.

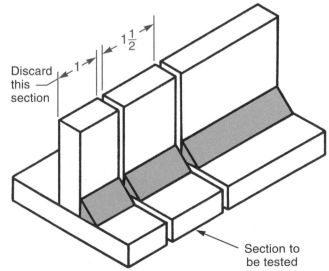

Figure 21-10.
Cutting the test section from a practice piece.

weave pattern for the second and third beads utilizes the force of the arc to wash molten metal onto the vertical surface. See **Figure 21-11.**

4. Clean and check the weld for appearance, size, and integrity. Also, check the plates for distortion.

5. To conserve material, run a similar weld on the other side of the joint.

6. Make additional practice pieces, but this time alternate the weld sequence so the weld builds equally on both sides of the joint. Compare the distortion of one of these practice pieces with your first piece using this welding technique.

7. When you have developed the skill to make a satisfactory weld, make up a test piece that has been welded on *one* side only. Test this weld in the same manner you tested previous welds.

8. Tilt one practice piece as shown in **Figure 21-12** and build up a fillet weld with the work in this position. Compare your results with previously made pieces.

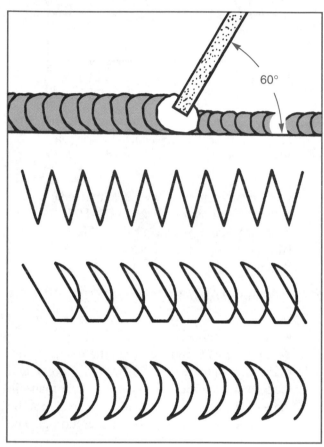

Figure 21-11.
Weave patterns for building up the weld.

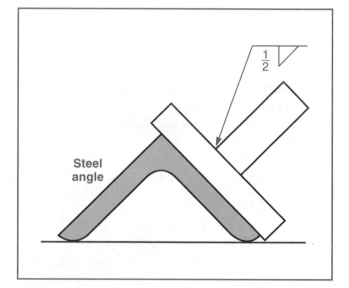

Figure 21-12.
Position for making the fillet weld with the work in a flat position.

Name _____ Score _____

Check Your Progress

1. For maximum strength, fillet welds should be made on _____ of the joint.

2. Multiple-pass fillet welds are best made with the following three types of electrodes:

   a. _____

   b. _____

   c. _____

3. Why is it often practical to tilt the pieces as shown in **Figure 21-12** when making a multiple-pass fillet weld?

   _____

4. Describe the differences between the two methods used for depositing multiple-pass fillet welds.

   _____

   _____

5. When using a weaving pattern to create a multiple-pass fillet weld, the second and third beads utilize the force of the arc to _____ molten metal onto the vertical surface.

Things to Do

1. Three different electrodes were recommended for making multiple-pass fillet welds. Prepare three similar practice pieces, each using a different type electrode. Weld only one side of the joint.
   Evaluate the practice pieces according to appearance, ease in welding, time required to make weld, problems with slag, undercutting, etc. Answer the following questions: Were there any differences? If so, which electrode gave the best results? How do your findings compare with the findings of your fellow students? Additional test results can be obtained by cutting a 1 1/2″ (38 mm) section from each practice piece and testing until the joint is broken or flattened.

2. In the space provided, sketch in the weld bead patterns for the two techniques recommended for running multiple-pass fillet welds.

Notes

Unit 22

Single V-Groove Butt Welds

After completing this unit, you will be able to:

○ Describe a situation when beveling the edges of plates is necessary.
○ Explain the purpose of an air carbon arc torch.
○ Make a single V-groove weld on a butt joint that is uniform, smooth, and evenly penetrates both plates.

Square groove butt joints can be used on plate up to 3/8″ (10 mm) thick if the weld can be made from both sides. However, there are many times when the weld can only be made from one side. In these situations, it is necessary to bevel the edges to form a 60°V along the joint on metal 1/4″ (6 mm) and thicker, **Figure 22-1.** Plate thicker than 5/8″ (16 mm) should be beveled on both sides, **Figure 22-2.**

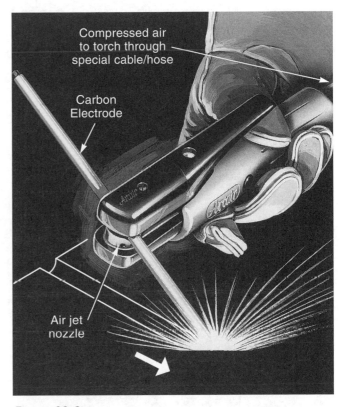

Figure 22-3.
Air carbon arc torch that will groove, bevel, cut, gouge, or pierce metal at a rapid rate. A blast of compressed air flushes the molten metal away. The joint needs no grinding or chipping when made with this tool.

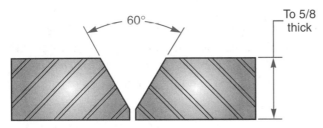

Figure 22-1.
Joint preparation for metal up to 5/8″ thick.

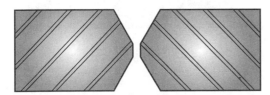

Figure 22-2.
Joint preparation for metal thicker than 5/8″.

The plates can be beveled with a grinder, an oxyacetylene cutting torch, a plasma torch, or an air carbon arc torch, **Figure 22-3.** The air carbon arc torch uses a carbon-graphite electrode to strike an arc on the metal to be gouged (cut away). The metal melts instantly and high velocity air jets come out of the torch heat to blast the metal away, **Figure 22-4.**

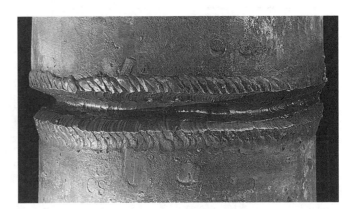

Figure 22-4.
Joint prepared with an air carbon arc torch. (Arcair Co.)

The edges can be beveled to a 1/8″ (3 mm) shoulder, sometimes called a land, or to a featheredge, **Figure 22-5.** The 1/8″ shoulder is preferred for economy and to prevent melt-through.

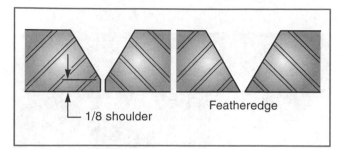

Figure 22-5.
Two types of beveled edges. The 1/8″ shoulder is preferred if no backing strip is used.

Study the plans in **Figure 22-6** carefully and secure the necessary material. Bevel the edges. Use 1/8″ (3 mm) diameter E6012 or E7018 electrodes for the root and cover passes. Larger 5/32″ diameter electrodes can be used for the cover passes to improve economy. **Figure 22-7** shows how the three beads are deposited.

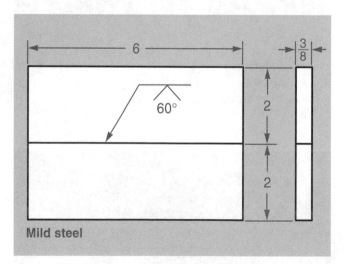

Figure 22-6.
Plans for the practice piece.

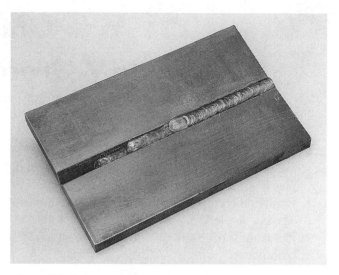

Figure 22-7.
Partially completed single V-groove butt weld showing how the three beads are run.

Procedure:

1. Make the usual preparations to weld. Be sure to wear safety glasses.
2. Position the plates and tack weld them in alignment. Space the joint as shown in **Figure 22-8.**
3. Run a bead along the bottom of the V. Be sure the bead penetrates both sides of the joint evenly. Weld on the back side of the joint indicates full penetration. This first pass is called a ***root pass.***
4. Clean the first bead thoroughly.

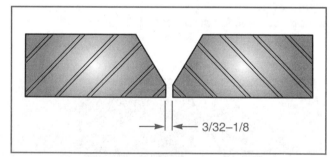

Figure 22-8.
Recommended spacing of the two plates.

5. Run additional beads until the weld deposit is built up slightly above the surface of the plates, **Figure 22-9.** Do not forget to clean each weld thoroughly before starting the next pass.

Note: It may be necessary to weave the last pass (called a *cover pass*) to ensure that the weld penetrates each side of the V evenly. When running these beads, maintain an electrode speed just fast enough to keep it slightly ahead of the molten pool.

6. Clean and examine the weld. It should be uniform, smooth, and penetrate evenly into both plates.
7. Cut out a test section as shown in **Figure 22-10.** Bend the piece until it breaks or is flattened.

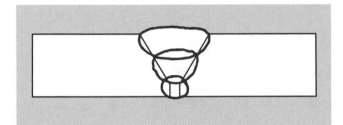

Figure 22-9.
Completed weld.

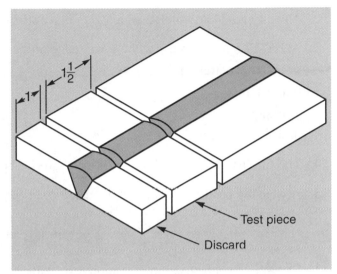

Figure 22-10.
Cutting test specimen from practice piece.

Name _____   Score _____

Check Your Progress

1. When square groove butt joints are made from one side of thick steel, the edges must be _____ to a 60° angle.

2. The air carbon arc torch uses a _____ electrode to strike an arc on metal to be cut away.

3. A 1/8″ (3 mm) shoulder at the lower edge of the V-groove joint is for:

   a. better penetration.

   b. economy.

   c. easier welding.

   d. a stronger joint.

4. The first pass along the bottom of the V-groove butt joint is called _____.

5. A weaving bead deposited last to ensure even penetration on each side of the V is called a _____.

Things to Do

1. Explain why a beveled joint is more expensive to produce than a square grooved butt joint.

2. There are several other ways of making single-bevel grooved joints. Make additional practice pieces using the techniques that follow. Compare these welded joints with your best previous practice piece.

Test Specimen No. 1.

   a. Set up and prepare your metal for welding as explained in the previously described sequence.

   b. Run the *root pass* with an E6011 electrode.

   c. Make the remaining passes with E6024, E6027, or E7018 electrodes.

Test Specimen No. 2.

   a. Position and tack weld the plates to hold them in place. The recommended spacing for this test piece is shown in **Figure 22-11.**

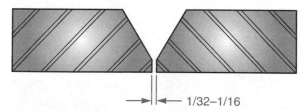

Figure 22-11.
Spacing of plates for Test Specimen No. 2.

b. Whenever possible, it is recommended that the weld be made from both sides. Run the welds in the sequence shown in **Figure 22-12.** Use an E6011 electrode for the first pass and an E6024, E6027, or E7018 for the other beads, **Figure 22-13.** If necessary, use an extra bead to build up the weld deposit until it is slightly above the surface of the plate. Do this instead of making the last pass heavier. There is a tendency for heavy welds to have slag and gas inclusions.

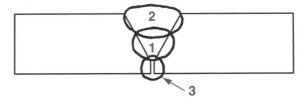

Figure 22-12.
Sequence recommended for running beads.

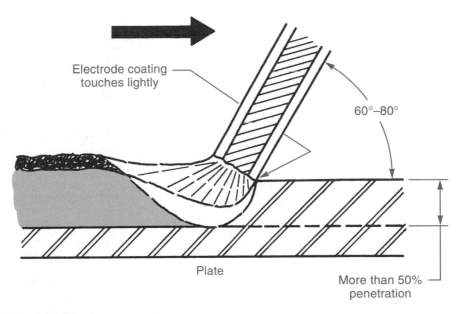

Figure 22-13.
Employ a drag technique when using E6024 or E6027 rod. Tip the electrode as shown.
Weld with the electrode lightly dragging on the work to force the molten metal out
from under the electrode tip and provide adequate penetration.

3. Remove a test section from Specimens 1 and 2 as shown in **Figure 22-10.** Bend them until broken or flattened. Examine the welds carefully and compare them using the standards of acceptability on page 97.

Mark the first test specimen No. 1, the second No. 2, and the third No. 3 so you will know what welding technique was used on each piece.

Notes

Unit 23
Corner Joint Fillet Welds

After completing this unit, you will be able to:
○ Explain the use of both the full-open corner joint fillet weld and the half-open corner joint fillet weld.
○ Make a full-open corner joint fillet weld that is uniform, evenly penetrates both plates, and does not create distortion.
○ Complete a half-open corner joint fillet weld that is uniform, evenly penetrates both plates, and does not create distortion.

The corner joint fillet weld is made in much the same way as the V-groove butt weld in that the plates form a V at the joint. Maximum strength is obtained with the *full-open corner joint fillet weld,* **Figure 23-1.** The *half-open corner joint fillet weld,* **Figure 23-2,** is employed when full strength is not an important factor or the joint need only be liquid tight. It can be welded faster and the fit of the joint is not as critical as that needed for a full-open corner joint fillet weld. However, additional strength can be obtained on the half-open corner joint by running an additional fillet weld on the inside of the joint, **Figure 23-3.**

The plans in **Figure 23-4** will provide all the necessary information. If the material is 3/8″ or thicker, the weld can be made with 1/4″ (6 mm) diameter E6024 electrode. For material less than 3/8″ thick, use 1/8″ diameter E6013 or E7018 electrode.

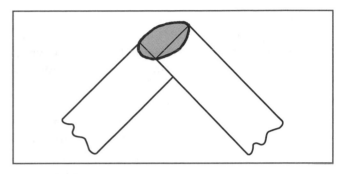

Figure 23-2.
Half-open corner joint fillet weld.

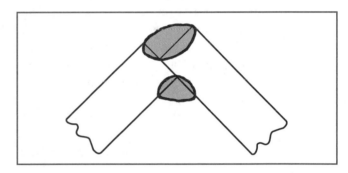

Figure 23-3.
Additional strength can be obtained on a half-open corner joint by running a fillet weld on the inside of the joint.

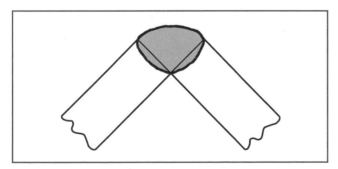

Figure 23-1.
Full-open corner joint fillet weld.

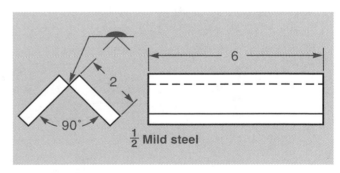

Figure 23-4.
Plans for this job.

Figure 23-5 shows the layers of beads as they are deposited in this weld.

Procedure: *Full-open corner joint fillet weld*

1. Make the usual preparations for welding. Be sure to wear safety glasses.
2. Position the plates and tack weld them in alignment, **Figure 23-6.**
3. Make the first pass. As the plates come to a featheredge at the joint, there is a possibility of burn-through. If burn-through is encountered, weave the electrode from one side of the joint to the other.
4. Clean and inspect the weld.
5. The second pass is made in a similar manner. However, when weaving the electrode from one side to the other, touch the top corner of one plate and stop the weave long enough that the edge is just ready to melt, then move across to the other edge and repeat the sequence until the weld is completed.
6. Clean and examine the weld for fusion at the root, even penetration in both plates, overhang, burn-through, uniform appearance, etc. Also, check whether the angle is 90°. If distortion has pulled the plates out of alignment, make allowances for the problem when you tack weld additional practice pieces.
7. Remove a 1 1/2″ (38 mm) section from the practice piece. Test until the weld breaks or the plates are flattened. Check the weld for fusion and penetration.

Figure 23-5.
Partially completed joint showing how the beads are deposited.

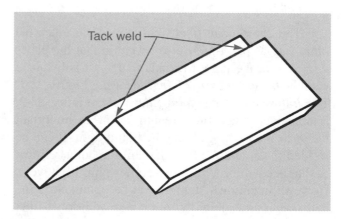

Figure 23-6.
Tack weld the plates to hold them in position for welding.

Name _____   Score _____

Check Your Progress

1. The half-open corner joint weld is used when:
   a. _____
   b. _____

2. Additional strength can be obtained on the half-open corner joint weld by running a fillet weld on the _____ of the joint.

3. If burn-through is encountered, what should you do to overcome this problem? _____
   _____
   _____

4. List five things to look for when examining the finished weld as to quality.
   a. _____
   b. _____
   c. _____
   d. _____
   e. _____

5. Maximum strength is obtained with the _____ corner joint weld.

Things to Do

1. What problems were encountered when making the full-open corner joint weld? What precautions can be taken to correct or minimize the problems?

Problems	How to correct

2. Contact a commercial welder and ask what precautions he would take when welding a container that holds a flammable liquid. If possible, have him relate to the class some of the outstanding welding problems he has encountered in his career and how he was able to solve them.

3. Set up and weld a *half-open corner joint weld.*

4. Do not destroy all your practice pieces. Use them to practice running a fillet weld on an inside corner.

Unit 24
Welding Round Stock

After completing this unit, you will be able to:

○ Explain the steps available to prevent distortion when welding round stock.

○ Use a butt joint to weld two pieces of round stock.

○ Successfully weld a piece of round stock to a heavier base material.

Welding plays a very important part in modern manufacturing. The process is employed to fabricate many stock metal shapes, **Figure 24-1,** into finished products.

When round stock (shafts, axles, rods, etc.) is welded, the butt joint is most frequently employed. The ends of the rounds are beveled by grinding, cutting, or machining, as shown in **Figure 24-2.** The angles must be the same size on both pieces. Leave a shoulder or land across the center.

Hold the pieces in alignment by clamping them in a vise or by placing them in a trough made from angle iron, **Figure 24-3,** and tack weld. To keep distortion to a minimum, alternate the welding pass

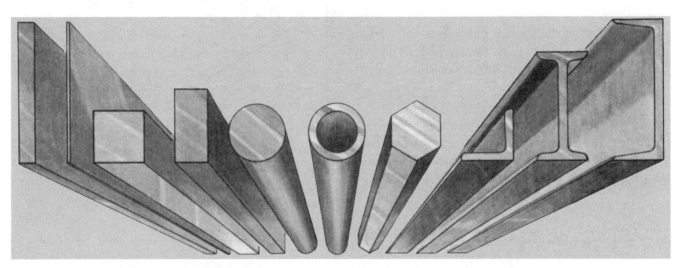

Figure 24-1.
Welding is often employed to fabricate the many shapes and sizes of metal into finished products.

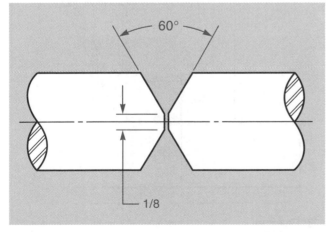

Figure 24-2.
When welding round stock, shape the ends as shown.

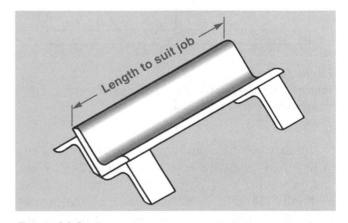

Figure 24-3.
A trough made of angle iron can be used to hold round stock in alignment while it is being welded. Weld short sections of smaller angle iron on the sides to hold the trough in a level position.

sequence so that the weld is built up equally on both sides of the joint, **Figure 24-4.** This will allow the forces that cause distortion to balance out each other. However, if distortion does occur, the shaft can be straightened on a hydraulic press.

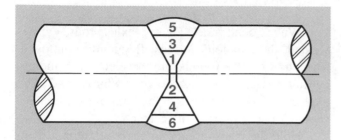

Figure 24-4.
Distortion can be kept to a minimum by alternating the welding sequence so that the weld is built up evenly on both sides of the joint.

Tubing or small diameter pipe can be joined with a straight end-to-end butt joint, **Figure 24-5,** a butt joint with a backing ring, **Figure 24-6,** or consumable insert.

 **Warning:** Be sure that there is adequate ventilation when welding galvanized (zinc coated) pipe. Fumes created during this process may be harmful to your health.

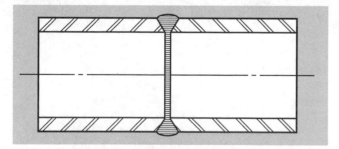

Figure 24-5.
Tubing or small diameter pipe can be joined with a straight end-to-end butt joint.

It may also be necessary, at times, to weld tubing or pipe to a heavier base section, **Figure 24-7.** When welding such a joint, position the electrode as shown in **Figure 24-8,** to direct more heat to the heavier metal. This will prevent the tubing from burning through. Tack welds should be placed 90° apart to secure the tube or pipe to the base section. The tacks will also prevent the pipe or tube from lifting off the base's section opposite the weld.

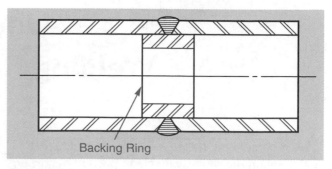

Figure 24-6.
A butt joint with a backing ring may also be used when welding tubing and pipe.

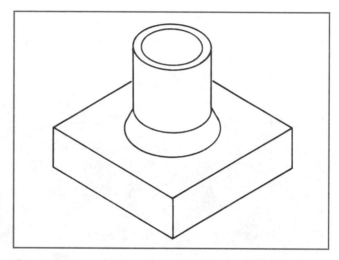

Figure 24-7.
At times, the welder must weld tubing or pipe to a heavier base section.

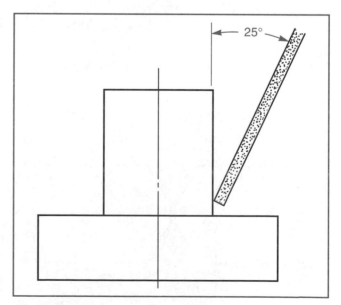

Figure 24-8.
Position of electrode when welding tubing or pipe to a heavy base so that more heat will be directed to the heavy base. This will reduce the possibility of burning through the lighter tubing.

Name _____ Score _____

Check Your Progress

1. The butt joint is most frequently employed when welding round stock, such as:
   a. _____
   b. _____
   c. _____

2. Tubing or small diameter pipe can be joined with a straight _____ butt joint.

3. Round pieces may be held in alignment for welding by clamping in a vise or placing them in a _____ made from angle iron.

4. The ends of round stock should be _____ by grinding before beginning the weld.

Things to Do

1. Weld two sections of 1″ (25 mm) diameter rod together. Draw plans of the job before welding.

2. Weld two sections of thick wall tubing together. Draw plans of the job before welding.

3. Weld a section of tubing to a heavier base plate. Draw plans of the job before welding.

4. Draw in the welding symbol for the job below.

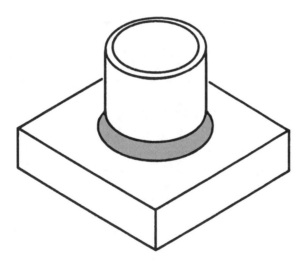
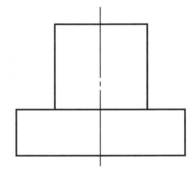

Notes

Unit 25
Welding in the Horizontal Position

After completing this unit, you will be able to:
- ○ Prevent the weld from sagging by properly controlling the arc.
- ○ Complete a horizontal square groove weld on a butt joint that evenly penetrates both plates to at least fifty percent of the metal thickness.
- ○ Make a horizontal V-groove weld on a butt joint that is smooth and covers the entire area of the joint.

Whenever possible, welding should be done in the flat, or downhand, position. However, because of the type or size of the job, welding cannot always be done in this position. To handle these jobs, you must learn to weld in the horizontal, vertical, and overhead positions. This unit is concerned with welding in the horizontal position, **Figure 25-1.**

Running Stringer Beads in the Horizontal Position

Set a piece of mild steel that is 4″ × 6″ × 3/16″ or 1/4″ (102 mm × 152 mm × 5 mm or 6 mm) thick in a vertical position, **Figure 25-2.** It can be clamped in a

fixture or tack welded to a steel plate to hold it steady. Make the beads with 1/8″ (3 mm) diameter E6010 or E6011 electrodes.

Practice running stringer beads. Lean the electrode as shown in **Figure 25-3.** Strike the arc and keep it short. If there is a tendency to undercut the plate at

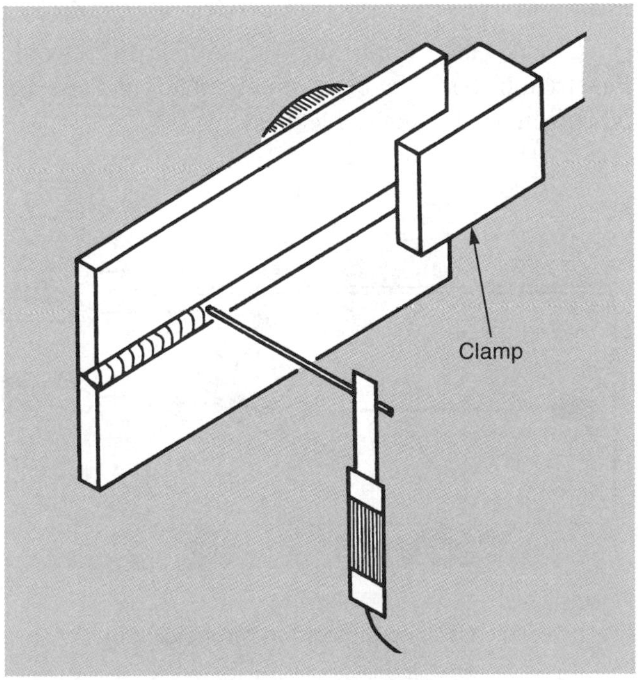

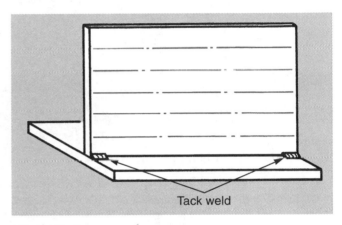

Figure 25-2.
Practice piece for horizontal welding.

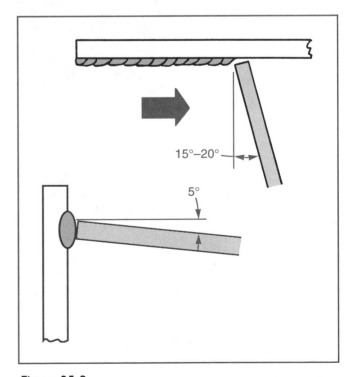

Figure 25-3.
Recommended positions of the electrode when making horizontal welds.

Figure 25-1.
The horizontal welding position.

the top of the bead, **Figure 25-4,** decrease the electrode angle until you get a normal bead, **Figure 25-5.** The upward position of the electrode permits the arc force to hold the molten metal in position until it freezes. Using slightly less amperage may be helpful in producing an acceptable bead, and weaving the electrode will help distribute the heat more evenly, **Figure 25-6.** A whip technique can also be used with E6010 or E6011 electrode to control the heat in the weld joint, **Figure 25-7.**

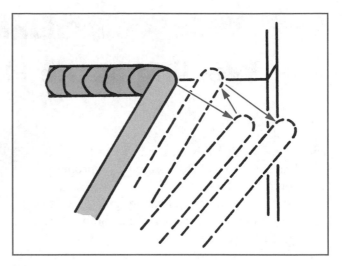

Figure 25-7.
A slight "whipping" motion of the electrode in and out of the molten weld pool helps control the heat in the joint.

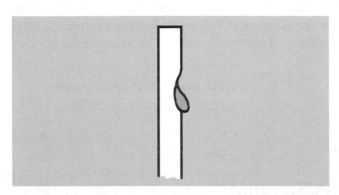

Figure 25-4.
This weld has sagged. Note the undercutting.

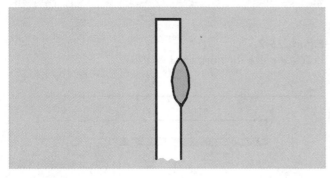

Figure 25-5.
Normal weld bead.

The secret for successful horizontal welding is to keep the weld pool small so that there is little tendency for the molten steel to sag. Continue running practice beads from left-to-right and right-to-left until you can run uniform beads with no overlapping or undercutting.

Making a Horizontal Butt Weld

Secure two pieces of mild steel 3″ × 6″ × 3/16″ or 1/4″ (76 mm × 152 mm × 5 mm or 6 mm) thick and tack weld them together. Clamp the plates so the joint is in a horizontal position, **Figure 25-8,** and make the weld.

Break the joint and examine the weld. Penetration should be even in both plates and at least 50 percent of the metal thickness.

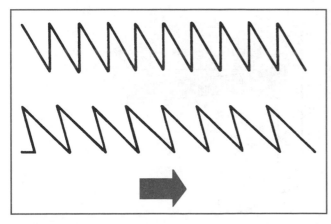

Figure 25-6.
Weaving patterns that will help distribute the heat more evenly.

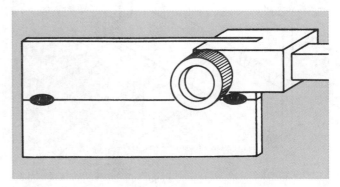

Figure 25-8.
Practice piece clamped in position for welding.

Making a Horizontal V-groove Butt Weld

Cut two pieces of mild steel 2″ × 6″ × 3/8″ (53 mm × 152 mm × 10 mm) thick. Bevel the edge as shown in **Figure 25-9.** The joint can be welded satisfactorily from one side. However, for complete fusion, welding from both sides is recommended. Use 1/8″ (3 mm) diameter E6010 or E6011 electrodes.

Tack weld the plates together and clamp so that the joint is in a horizontal position. Deposit the root bead. Remove the slag and make the second pass. The last or wash pass is made with a weaving motion to cover the entire area of the joint and should produce a smooth finish.

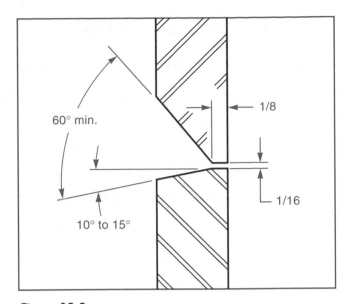

Figure 25-9.
Groove design for horizontal welding. The bottom of the joint serves as a shelf and helps prevent the molten metal from running out of the joint.

Name _____   Score _____

Check Your Progress

1. Whenever possible, welding should be done in a flat, or _____, position.

2. The secret for successful horizontal welding is to keep the puddle _____.

3. When beginning a horizontal weld, strike the arc and keep it _____.

4. More even distribution of heat during welding can be accomplished by _____ the electrode.

5. If there is a tendency to undercut the plate at the top of the bead, you should:
   a. tilt the plate being welded.
   b. increase the amperage.
   c. decrease the electrode angle.
   d. hold a longer arc.

Things to Do

1. Practice running stringer beads from left-to-right and right-to-left until you can make welds that are uniform, with no undercutting or overlapping.

2. Prepare additional horizontal butt welds. Practice making them until a satisfactory joint can be made without difficulty.

3. Practice making horizontal V-groove butt welds until you become proficient in welding joints of this type.

4. The backing strip or bar is a strip of steel that is placed against the back of the weld. It is frequently used to make it easier to weld joints with a great deal of mismatch. During the root pass, the backing strip is fused to the weld. Research when and how the backing strip is used. In addition, see if you can learn what consumable-insert backings are and how and when they are used.

Unit 26
Welding in the Vertical Position

After completing this unit, you will be able to:
○ Explain the basic differences between vertical-down welding and vertical-up welding.
○ Create uniform vertical-up and vertical-down welds.
○ Make both groove welds and fillet welds in the vertical position that are consistent and have the proper penetration.

A vertical weld is made either upward or downward. *Vertical-down welding,* **Figure 26-1,** is the easier of the two. The technique is best suited for welding thin material (less than 3/16″ [5 mm]) because penetration is shallow and reduces the possibility of burn-through.

The greater penetration of *vertical-up welding,* **Figure 26-2,** makes it better suited for vertical welding of thicker material. Some welding codes may also require that vertical joints be welded with a vertical-up progression.

Regardless of whether the dirction of travel is upward or downward, the electrode is held at right angles to the face of the weld and inclined downward (electrode holder below arc). See **Figure 26-3.**

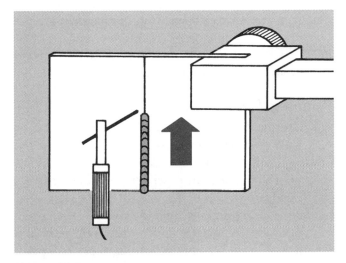

Figure 26-2.
Vertical-up welding.

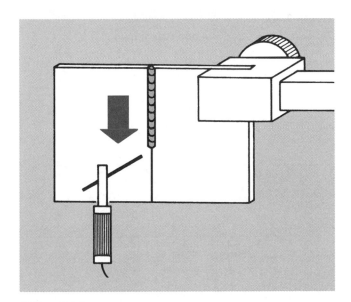

Figure 26-1.
Vertical-down welding.

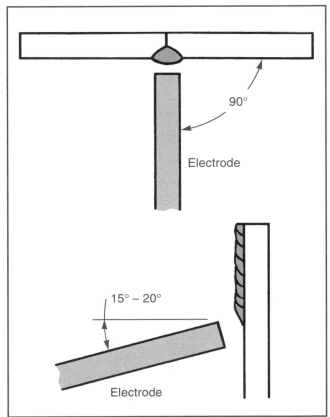

Figure 26-3.
Recommended position for electrode when welding work in the vertical position.

Running Vertical-Down Stringer Beads

Clamp or tack weld a piece of mild steel 4″ × 6″ × 3/16″ (102 mm × 152 mm × 5 mm) in position. Use 1/8″ (3 mm) diameter E6010 or E6011 electrodes.

Strike the arc and proceed to move the electrode downward just fast enough to keep the molten metal and slag from running ahead of the crater. This will require some practice. Use a short to medium arc.

Practice running beads (both sides of the practice piece may be used) until you can maintain an even speed that will produce a uniform bead.

Running Vertical-Up Stringer Beads

Use the same size practice piece and clamp or tack weld it in position. Strike the arc and move the electrode upward with a slight weaving or rocking motion. The slight motion will move the electrode's tip in and out of the crater and allow the deposited metal to solidify and build up the weld.

Hold the arc length short in the weld pool and lengthen it as it is removed from the weld pool, **Figure 26-4.** The "whip" technique is used with E6010 and E6011 fast-freeze electrodes. Practice running vertical-up beads until you can produce acceptable welds. For E7018 low hydrogen electrodes, use the weave patterns used for fillet welds, which are discussed later in this chapter.

Making Groove Welds in the Vertical Position

Secure two pieces of mild steel plate 3″ × 6″ × 3/16″ (76 mm × 152 mm × 5 mm) thick. Use 1/8″ (3 mm) diameter E6010 or E6011 electrodes.

Tack weld the long edges and clamp the plates in position, **Figure 26-5.** Strike the arc and move the electrode down. Penetration should be 50% of plate thickness.

You would normally make the weld from both sides to secure complete fusion. However, the practice piece is to be broken to check the weld for penetration and evenness of fusion in both pieces.

To make a vertical-up groove weld on a butt joint, secure two pieces of mild steel plate 3″ × 6″ × 3/8″ (76 mm × 152 mm × 10 mm) thick. Prepare the joint as shown in **Figure 26-6.** Tack weld and clamp the plates in position.

Strike the arc and make the root pass as practiced. Clean off the slag. Deposit additional beads until the joint is completed. The last pass should be a wash pass. Test the joint until it breaks or is flattened. Check for good weld characteristics.

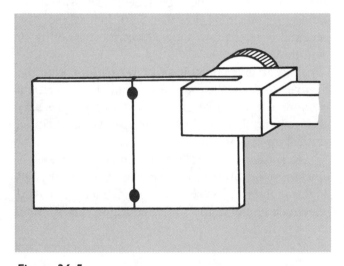

Figure 26-5.
Tack weld to keep distortion to a minimum.

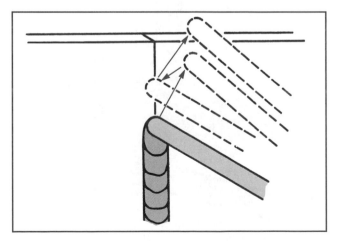

Figure 26-4.
The slight "whipping" motion of the electrode in and out of the molten weld pool allows the deposited metal to solidify and build up the weld.

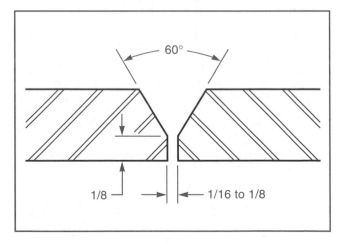

Figure 26-6.
Joint preparation for vertical welding.

Making Fillet Welds in the Vertical Position

Secure 3/8"–1/2" (9.52 mm–13 mm) thick mild steel of the sizes shown in **Figure 26-7.** Tack weld and clamp them in position.

To make a fillet weld in a vertical upward position, the first bead is run in a weaving motion similar to that shown in **Figure 26-8.** The pattern spreads the heat evenly to control the size of the weld pool and allows good removal of impurities.

A smoother bead can be formed with the weave pattern shown in **Figure 26-9;** however, greater care must be taken to prevent slag entrapment in the center of the pattern.

To enlarge the weld, the pattern in **Figure 26-10** is recommended. Slight upward movement is used to tie the weld into the base plates while keeping the possibility of undercutting to a minimum.

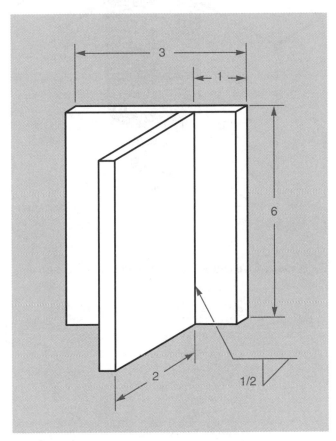

Figure 26-7.
Size and shape of practice piece for making fillet welds in a vertical position.

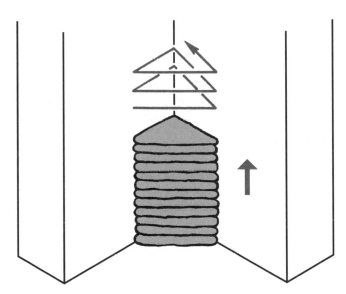

Figure 26-9.
Alternate weave pattern for making a vertical-up fillet weld. Care must be taken to prevent slag entrapment in the center of the pattern.

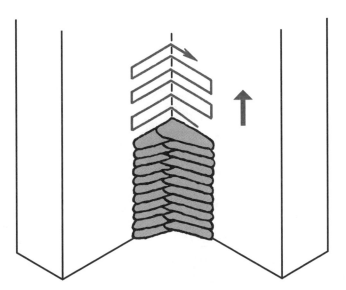

Figure 26-8.
Weave pattern for building up a vertical-up fillet weld.

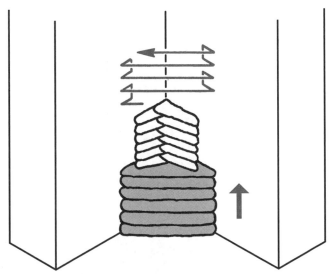

Figure 26-10.
Weave pattern for enlarging a vertical-up fillet weld.

Practice the weld pattern until satisfactory welds can be made with consistency. Test your best practice piece for weld integrity by using the same method employed to test previous fillet welds.

It is not recommended that plate thicker than 3/16″ (5 mm) be welded in a vertical downward direction. However, thicker metal can be welded if a series of beads only two or three times the width of the electrode are run. The beads are made using a slight zigzag weave pattern, **Figure 26-11.** Since a small bead is employed, it may be necessary to run several beads to build the weld up to the required size.

The problem of porosity and slag inclusion, common to this method of welding, can be controlled by limiting the heat input and moving the electrode only fast enough to keep molten metal and slag from running ahead of the crater.

Since this welding technique is quite difficult, much practice will be required before a satisfactory weld can be made with consistency. Test the weld by using the same method used to check the soundness of earlier fillet welds.

Note: Vertical-down welding is generally used only on sheet metal and auto body work where burn-through can cause problems.

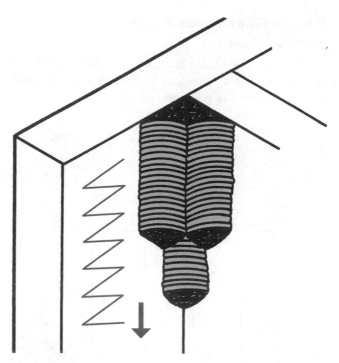

Figure 26-11.
Suggested weave pattern for building up a vertical-down fillet weld.

Name _____ Score _____

Check Your Progress

1. Vertical-up welding is better suited to thick material because it gives better _____.

2. In both vertical-up and vertical-down welding, the electrode is held at right angles to the face of the weld and inclined _____.

3. Penetration should be _____% of the plate thickness when making vertical butt welds.

4. Vertical-down welding is best suited when using thin material because:

   a. _____

   b. _____

5. The electrode should be moved downward just fast enough to keep the molten metal and slag from running _____ of the crater.

Things to Do

1. Considerable skill is required to make satisfactory welds in the vertical position. Constant practice is the only way this skill can be developed. Practice making several welds in the vertical position. Examine each practice weld carefully and evaluate weld soundness. Note flaws or welding difficulties and make every effort to correct any problems.

2. Visit a professional welder. Ask him for information about the problems you have encountered. If possible, have him show you the techniques he uses in vertical welding.

Notes

Unit 27
Welding in the Overhead Position

After completing this unit, you will be able to:
○ List five safety precautions recommended when welding in the overhead position.
○ Run a uniform bead with adequate penetration in the overhead position.
○ Make an acceptable fillet weld in the overhead position.

The overhead position, **Figure 27-1,** is one of the most difficult to master of the four welding positions. Not only are you required to stand in a tiring and awkward position, but you must also contend with working in a shower of molten metal and slag that is continually falling. Gravity makes it difficult to keep the molten weld pool from dropping. This causes problems because it is harder to get good weld penetration and a bead that is uniform.

Warning: Since there is danger from falling molten metal, it is important that you dress properly. Protect your head with a cap, button your pockets, wear leather sleeves and shoulder covers, and make sure your pant legs cover the tops of your shoes.

Running Stringer Beads in the Overhead Position

Clamp a piece of 3/16″ or 1/4″ (5 mm or 6 mm) mild steel plate in position, **Figure 27-2.** Use 1/8″ (3 mm) diameter E6010 electrode, and maintain the electrode angles shown in **Figure 27-3.** For E7018

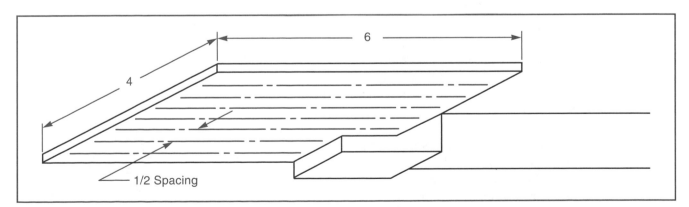

Figure 27-1.
Welding in the overhead position.

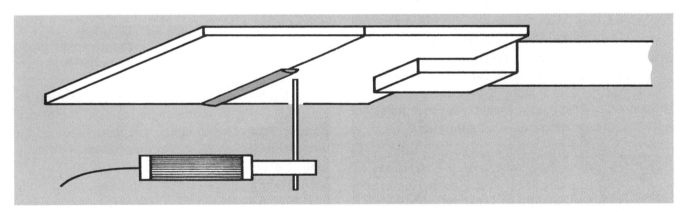

Figure 27-2.
Size of steel plate and spacing of welds for practice piece.

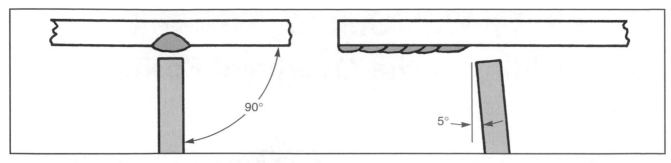

Figure 27-3.
Recommended electrode angles for overhead welding.

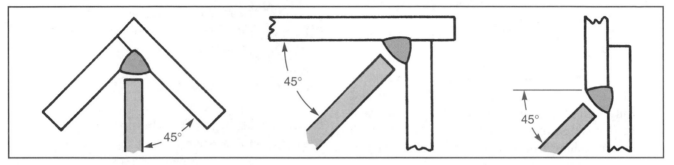

Figure 27-4.
For overhead lap and fillet welding, position the electrode so it splits the angle between the plates.

1/8″ diameter electrodes, use a slightly larger drag angle (about 10°). A short side-to-side weave helps prevent sticking.

Overhead welding can be made more comfortable. When standing, drape the cable over your shoulder. Drape it over your knees if you can sit to weld.

It will be less fatiguing if you support the electrode holder with both hands and, whenever possible, try to steady one elbow on something solid.

> **Warning:** Grip the holder so that the molten metal rolls off your gloves. Keep your knuckles up and your palm down.

Strike the arc and keep it short so that the weld pool does not get too large. If there is some tendency for the molten metal to drip, reduce amperage slightly. Often, just a very small change is all that is needed to keep the molten metal where you want it.

Move the electrode at such a speed that the force of the arc will hold the molten metal in position until it freezes.

You will find that you cannot weld as fast in the overhead position as you did in other positions. It is also more difficult to get a sound weld, because the impurities do not float to the surface of the weld pool.

Practice running beads until you develop the skill to control the molten pool. Then, work on getting adequate penetration and a uniform bead.

Welding Other Joints in the Overhead Position

To make butt, lap, and tee joints in an overhead position, it is better to run a series of small beads rather than a single large bead. Clean each weld thoroughly before depositing the next bead.

For overhead lap and tee joints, position the electrode so it splits the angle between the plates, **Figure 27-4.** Use a slight circular or triangular weave to agitate the weld pool. This will help remove slag and impurities from the pool.

When running the remaining beads, position the electrode so most of the heat is directed into the upper surface of the joint, **Figure 27-5.**

Practice making overhead position welds until the welds you produce are satisfactory. Test and compare them with welds made in other positions.

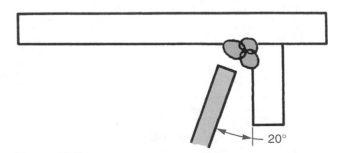

Figure 27-5.
To minimize undercutting, position the electrode so most of the heat is directed into the upper surface of the joint.

Name _____   Score _____

Check Your Progress

1. List five important factors concerning proper dress for welding in the overhead position.

   a. _____

   b. _____

   c. _____

   d. _____

   e. _____

2. It is difficult to keep the molten pool from dropping due to _____.

3. It is harder to get good _____ and a _____ bead when doing overhead welding.

4. Strike the arc and keep it _____ so that the weld pool does not get too large.

5. It is more difficult to get a sound weld in the overhead position because impurities do not _____ to the surface of the weld pool.

Things to Do

1. Contact a professional welder and ask about techniques that have been developed to produce acceptable welds in the overhead position.

2. Identify applications where overhead welds were made. The building you are in probably contains many such welds. Can you identify them?

3. Much practice is required to develop the skill to make overhead welds. Practice running stringer beads, making butt, lap, and fillet welds until you can make sound welds that are uniform in appearance.

Notes

Unit 28
Cutting with the Arc

After completing this unit, you will be able to:
○ List some of the additional safety precautions that should be observed when cutting with an arc.
○ Cut a piece of metal with an arc.
○ Create a hole in a piece of metal with an arc.

Cutting with an arc leaves something to be desired because it does not make a smooth precision cut, **Figure 28-1.** However, it is quick and often the most expedient way to make holes and to cut bolts, shafts, plates, or angles.

The intense heat generated by the arc can melt most metals. The force of the arc is used to propel the molten metal away.

Warning: Work only in a well-ventilated area and avoid fumes from galvanized steel, brass, lead, and paint. Also, the sparks produced are thrown a considerable distance; every effort should be made to remove any flammable materials from the area.

Since cutting with the arc is a burning process, it generates more fumes than welding.

The sparks can cause you much discomfort if you are not dressed properly for the job.

Warning: Be sure that you do not wear pants with cuffs. Have the bottom of your pants covering the top of your shoes or boots.

How to Cut with the Arc

1. Make cuts with a 1/8" (3 mm) diameter E6010 or E6011 electrode. Set machine to 140–180 amperes. Larger electrodes can be used if the machine is capable of higher settings.
2. Prepare the metal for cutting by positioning it so that the section to be cut projects beyond the edge of the table, **Figure 28-2.** Place a container of sand below the cut to catch the molten metal. In addition, keep the cables well clear of the molten metal.

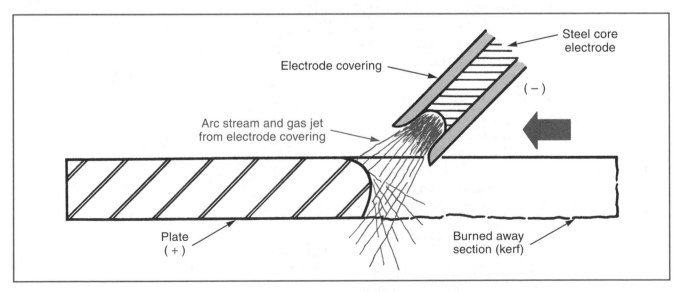

Steel core electrode

Electrode covering

(–)

Arc stream and gas jet from electrode covering

Plate (+)

Burned away section (kerf)

Figure 28-1.
Cutting with a steel electrode. Most cutting electrodes use a covering that disintegrates at a slower rate than the metal core of the electrode. A deep recess is formed in the arc end of the electrodes and produces a jet action that tends to blow the molten metal away.

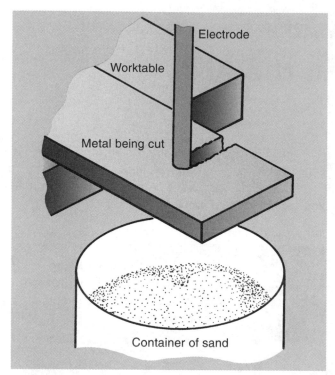

Figure 28-2.
Be sure the section to be cut clears the end of the bench. Position a container to catch the molten metal as it falls away.

3. Mark the area to be cut with soapstone or chalk. Strike the arc on the edge of the metal.
4. Hold a long arc until the metal starts to melt. Then shorten the arc to force the molten metal away. Continue this sequence (long arc, then short arc) as you cut across the plate, **Figure 28-3.**
5. Round stock can be cut in much the same way. Make the cuts from the side so the molten metal will drop clear of the work, **Figure 28-4.** Cut to the center of the work piece; then start a second cut from the other side of the piece.

How to Pierce or Cut a Hole with an Arc

It is also possible to burn openings with the arc. Remember, this is not a way to make an exact size hole, but the holes produced are satisfactory for some applications.
1. Holes can be made in plate thinner than 3/8" (10 mm) by striking the arc and holding it until the metal becomes molten.
2. Feed the electrode into the molten pool until the plate is pierced.
3. Larger holes can be made by feeding the electrode into the molten metal in a circular motion, **Figure 28-5.** Continue this motion until the hole is completed. Do not forget to use a container of sand to catch the molten metal.

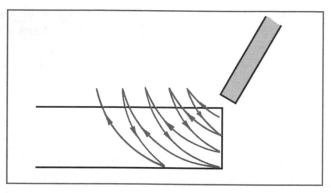

Figure 28-3.
Use this routine (long arc, then short arc) to burn the metal away.

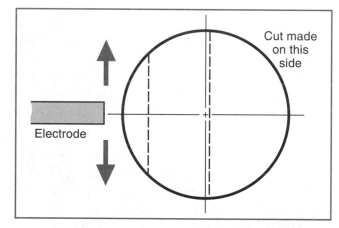

Figure 28-4.
Technique recommended to cut round stock. It permits the molten metal to drop clear rather than forming "icicles" that must be removed with a chisel or on the grinder.

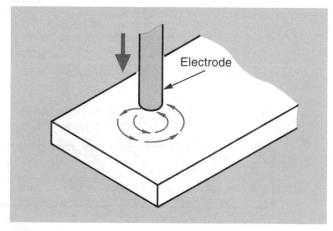

Figure 28-5.
Use a rotary motion to burn large diameter holes.

Note: It is easier to burn holes in metal thicker than 3/8" (10 mm) if the plate is placed on its side to permit the molten metal to run out and drop clear of the work.

Name _____ Score _____

Check Your Progress

1. Cutting with an arc is often a quick way to cut the following:

   a. _____

   b. _____

   c. _____

   d. _____

   e. _____

2. It is easier to burn holes in metal over 3/8″ (10 mm) thick if the plate is placed _____.

3. A container filled with _____ works well to catch the molten metal from arc cutting.

4. Cutting with an arc is a burning process and it generates more _____ than welding.

5. Large holes can be made in metal by feeding the electrode into the molten metal in a _____ motion.

6. Standard machine settings for cutting with a 1/8″ (3 mm) diameter electrode are from _____ to _____ amperes.

Things to Do

1. Practice making cuts on scrap pieces of steel, including steel bar, angle, plate, and rod.

2. After gaining experience by burning holes in scrap steel, burn 3/8″ (10 mm) diameter holes in 1/8″ (3 mm) steel plate.

3. Burn 1″ (25 mm) diameter holes in 1/4″ (6 mm) steel plate.

4. Lay out and cut a 2″ (51 mm) diameter circle in 1/4″ (6 mm) steel plate.

5. Prepare a list of safety precautions you should follow before starting to cut with the arc.

Notes

Unit 29
Welding Sheet Metal

After completing this unit, you will be able to:
○ Name the characteristics that define sheet metal.
○ List three techniques that prevent distortion when welding sheet metal.
○ Properly weld two pieces of sheet metal.

Sheet metal is usually thought of as a large, rectangular piece of metal less than 1/4" (6 mm) thick. Metal that is 1/4" or more in thickness is called *plate,* **Figure 29-1.**

A good welder can produce satisfactory welds in sheet metal after receiving some training in this highly specialized field.

When welding sheet metal, remember that the metal is thin. This makes it very important to use the correct heat (amperage). If the heat is too high, there will be burn-through. If there is not enough heat, incomplete fusion will result. The table in **Figure 29-2** will give you recommended amperage and the kind and size of electrode for different metal thickness.

The joints must be close fitting and carefully tack welded at several points to minimize distortion. This is a problem when welding thin metal.

It is good welding practice to back up the joint with a copper backing strip. The strip will help control burn-through and, because of its high conductivity, will help control distortion by carrying away excess heat. Weld metal will not stick to copper.

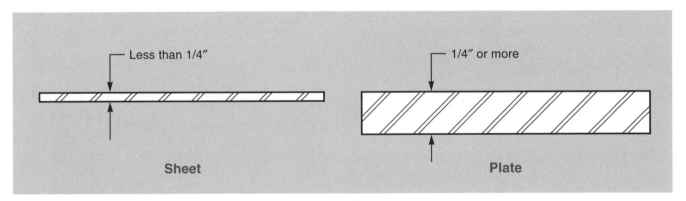

Figure 29-1.
Metal is considered sheet metal when it is thinner than 1/4" (6 mm).

Sheet Thickness Gage	Approximate Decimal Thickness	Approximate Fractional Thickness	Electrode Size	Electrode Type	Polarity	Amperage	Joint
16 ga.	.060	1/16	1/16	E6011	Neg.	30-60	Butt
14 ga.	.078	5/64	3/32	E6011	Neg.	40-60	Butt
12 ga.	.105	7/64	1/8	E6011	Neg.	80-100	Butt
10 ga.	.135	1/8	1/8	E6011	Neg.	90-100	Butt
16 ga.	.060	1/16	3/32	E6013	Neg.	75-100	Fillet
14 ga.	.078	5/64	1/8	E6012	Neg.	115-130	Fillet
12 ga.	.105	7/64	1/8	E6012	Neg.	120-140	Fillet
10 ga.	.135	1/8	5/32	E6012	Neg.	175-195	Fillet

Figure 29-2.
Recommended electrode size and amperage settings for welding sheet metal.

Heavy chill bars of steel or copper placed as shown in **Figure 29-3** will carry away excess heat and help hold the metal in position.

Another trick to keep distortion to a minimum is to tip the metal at an angle, **Figure 29-4,** and weld downward. This technique also decreases the tendency to burn through, especially if there is poor fit-up. In addition, it generally produces a better appearing weld and welding speed can be increased.

Maintain the same electrode position as recommended for making similar type welds in thicker material.

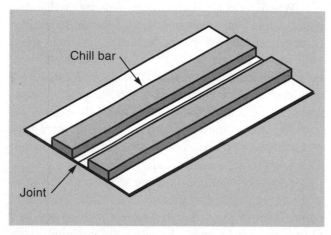

Figure 29-3.
Chill bars not only hold the sheet metal in place but also carry away excess heat.

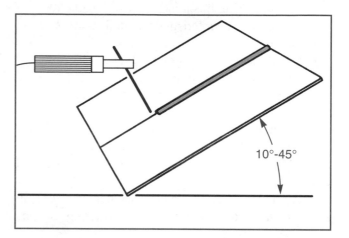

Figure 29-4.
Tipping the sheet metal at an angle improves the quality of the weld and makes it possible to increase welding speed.

Name _____ Score _____

Check Your Progress

1. Sheet metal is classified as being:
   a. less than 1/8″ (3 mm) thick.
   b. less than 1/4″ (6 mm) thick.
   c. less than 3/8″ (10 mm) thick.
   d. less than 1/2″ (13 mm) thick.

2. Incomplete _____ will result if there is not enough heat when welding sheet metal.

3. One of the major problems when welding sheet metal besides burn-through is _____.

4. Chill bars help carry away excess heat and help _____.

5. If you were welding 12-gage sheet metal and plan to make a fillet weld, you should select a _____ diameter electrode. The electrode type should be number _____.

Things to Do

1. Secure samples of sheet metal that has been joined by arc welding.

2. Visit a plant that specializes in welding sheet metal. Carefully observe how they weld different kinds of metal. What type fixtures are used to hold metal while it is welded?

3. Practice making various types of joints in sheet metal of different thicknesses. Compare welding metal of similar thicknesses flat and tipped at an angle.

Notes

Unit 30
Hardsurfacing

After completing this unit, you will be able to:
○ List the common uses of hardsurfacing.
○ Name the four methods of applying hardsurfacing.
○ Explain the benefits of each type of hardsurfacing.

Hardsurfacing (sometimes called *hardfacing*) involves using the heat of the arc to fuse a coating or layer of tough, wear-resistant or corrosion-resistant alloy to machine parts, **Figure 30-1.**

The process has made it possible to economically extend the working life of equipment by building up worn surfaces with material that will resist wear better than the base metal. While primarily used to rebuild parts, new components can be treated in the same manner.

Hardsurfacing is also employed to lower production costs. Less expensive low-alloy steels are used in many rebuilding jobs. Only the areas subjected to high wear are coated with the more expensive alloys required to stand up under severe service conditions.

Roadbuilding, grading, construction, and farming are but a few of the industries that make extensive use of hardsurfacing.

Service requirements for the parts to be hardsurfaced will determine the exact type of electrode that must be used. A different type of electrode is needed if the part is subject to abrasion (grading and farm equipment) than if the part is to withstand impact (rock crushers, mining equipment, and power shovel teeth).

It is not possible to cover the entire area of hardsurfacing in a unit as short as this. However, this unit will serve to introduce you to the possibilities of this welding technique.

Methods of Applying Hardsurfacing

On parts subject to very hard wear, the entire surface may be covered with a layer of hardsurfacing alloy, **Figure 30-2.**

Before welding, remove all rust, grease, and other foreign matter by grinding. Use the type of

Figure 30-1.
With use, many machined surfaces on this giant earthmover and bulldozer wear. Rather than discard the units, the worn surfaces are built up with hardsurfacing welding rod. This is often done before the equipment is used for the first time to extend its service life. (Waste Management)

electrode recommended by the electrode manufacturer for each specific application.

Beware of using an amperage setting that is too high. This will result in excessive mixing of the base metal and rod alloy, which can soften the weld deposit.

For many applications, it has been found that a pattern of stringer beads is more economical. Deposit the beads parallel with the flow of abrasive material on equipment that will handle heavy rocks, **Figure 30-3.** The hardsurfacing material will support the rocks while offering the least resistance to flow.

When the surface of the equipment is to be exposed to the abrasive action of sand, soil, and small stones, run the beads perpendicular to the flow, **Figure 30-4.** Dirt packs between the beads further protecting the base metal.

The diamond pattern, **Figure 30-5,** is often used. It is self cleaning because it prevents the dirt from packing between the beads. Generally, two layers of hard facing material should be applied to prevent dilution with the base metal and to develop the needed hardness.

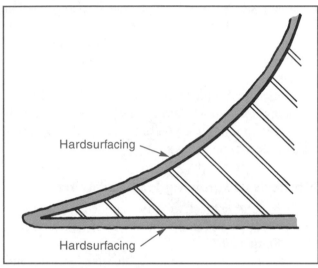

Figure 30-2.
When the part is to be subjected to very hard wear, the entire surface may be covered with a layer of hardsurfacing alloy.

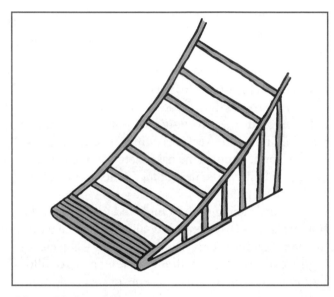

Figure 30-4.
Stringer beads of hardsurfacing alloy placed perpendicular to flow are recommended when the equipment will be exposed to the abrasive action of sand, soil, and small stones.

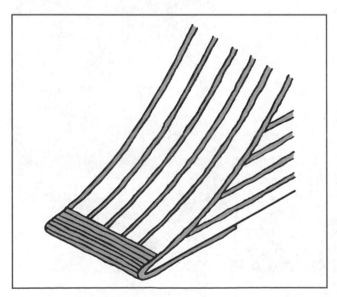

Figure 30-3.
Stringer beads of hardsurfacing alloy are more economical for some applications.

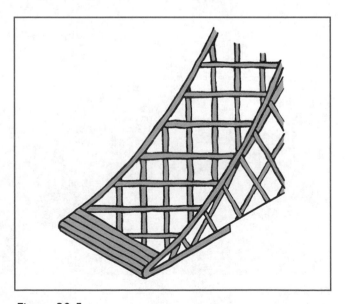

Figure 30-5.
The diamond pattern is self cleaning.

Name _____   Score _____

Check Your Progress

1. Four industries that make extensive use of hardsurfacing are:

    a. _____

    b. _____

    c. _____

    d. _____

2. Hardsurfacing is sometimes called _____.

3. The exact type of electrode to be selected for hardsurfacing is determined by _____.

4. For many applications, a pattern of _____ beads is more economical for hardsurfacing.

5. A diamond pattern of hardsurfacing is often used on equipment because it is _____.

Things to Do

1. Using electrode handbooks from a number of manufacturers, prepare a list of hardsurfacing electrodes suitable for the following jobs:

    a. Severe abrasion.

    b. Severe impact.

    c. Moderate abrasion.

2. Visit a heavy equipment dealer and talk with the shop supervisor on how hardsurfacing is used to repair and rebuild machinery.

3. Contact a farmer and volunteer to rebuild corn planter runners, plowshares, cultivator spikes, or some other equipment part with hardsurfacing.

4. Make a collection of literature on hardsurfacing.

5. A hardsurfacing paste is also available for use with the carbon arc. Research this method and prepare a report.

Notes

Unit 31
Carbon Arc Welding

After completing this unit, you will be able to:
- ○ Identify the materials that require the use of carbon arc welding.
- ○ Make a single carbon arc weld using a filler rod.
- ○ Complete an acceptable weld using a twin carbon arc torch.

Carbon arc welding (CAW) makes it possible to weld aluminum, brass, and copper almost as easily as ferrous (iron containing) metals can be metallic arc welded. The same welding machine is used.

In addition to welding, the flame-type heat produced by the carbon arc can be used for loosening rusted bolts, brazing, soldering, hardsurfacing, and heat treating.

With the single carbon arc torch, **Figure 31-1,** the heat is produced from an electric arc between the carbon electrode and the work. With the twin carbon arc torch, **Figure 31-2,** the arc is struck between the two carbon electrodes. No metal is transferred to the weld by the arc as in the metallic arc process. If extra metal is needed, it must be added to the joint by introducing a filler metal into the heat of the arc or into the molten puddle, similar to oxy-fuel welding.

Carbon arc welding electrodes are either pure graphite or baked carbon. Pure graphite electrodes are more expensive, but they last longer and can withstand higher heats. However, baked carbon electrodes are satisfactory for most jobs. The electrodes are copper coated for clean handling.

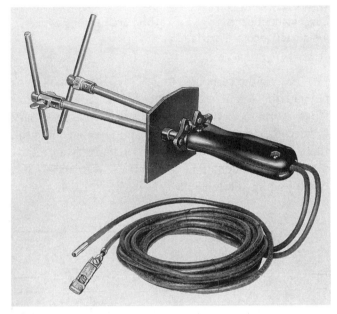

Figure 31-2.
Twin carbon arc torch. (Lincoln Electric Co.)

Twin Carbon Arc Welding Operations

In using the twin carbon arc torch, the distance between the carbons and the work controls the amount of heat going into the work. **Figure 31-3** indicates the carbon diameter and amperage settings for various metal thicknesses. An AC welder is used.

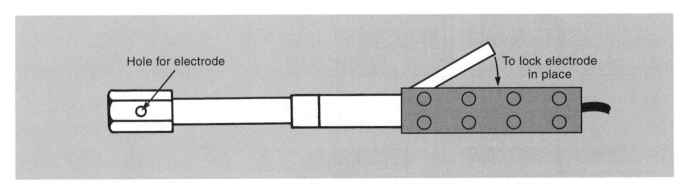

Figure 31-1.
Single carbon arc torch.

Carbon Diameter	Amperage Range	Metal Thickness
1/4	25–50	1/64 to 1/32
5/16	35–70	1/32 to 3/64
3/8	40–90	1/16 to 1/4
1/2	75–140	over 1/4

Figure 31-3.
Suggested ampere range for carbon arc carbons with AC welders. (Lincoln Electric Co.)

Extend the carbons 2 to 2 1/2″ (51 mm to 64 mm) beyond the torch jaws, **Figure 31-4.** Adjust the arc length (distance between carbons) as often as necessary to keep the arc going smoothly and to concentrate the flame in a small cone. As the carbon burns away, adjust the arc by manipulating the carbons with the thumb control on the torch handle.

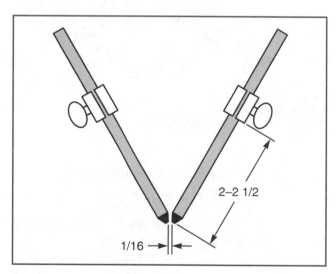

Figure 31-4.
Setting carbons on twin carbon arc torch.

Safety

The rays of the carbon arc torch are just as dangerous as those developed during conventional arc welding. A head shield and protective clothing are required.

Warning: Always turn the welder off when adjusting the carbons. This will prevent arcing should they touch during adjustment. The resulting flash could cause serious burns or damage to your eyes.

Joint Preparation

The thickness of the metal being welded will determine how the joint must be prepared, **Figure 31-5.** It is recommended that all welds be made in the flat position, even though this may require rotating the pieces.

Making the Weld

Follow this sequence for carbon arc welding:
1. Make the usual preparations for welding. Be sure to wear safety glasses and a number 14 shade lens in your welding helmet.
2. Set the welder to the current recommended in **Figure 31-3.**
3. Use a 1/8″ (3 mm) diameter filler rod (the same composition as the base metal). Hold it with your free hand.
4. Ignite the torch by bringing the carbons together and adjusting until the arc is going smoothly.
5. Keep the torch parallel to the joint as shown in **Figure 31-6.** Observe the welding process by looking between the two carbons.

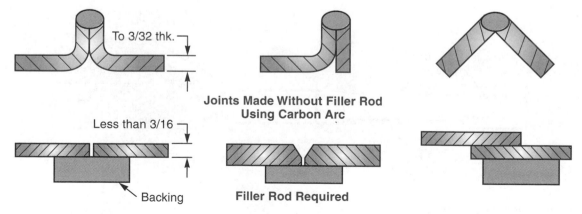

Figure 31-5.
Joint setup for carbon arc welding.

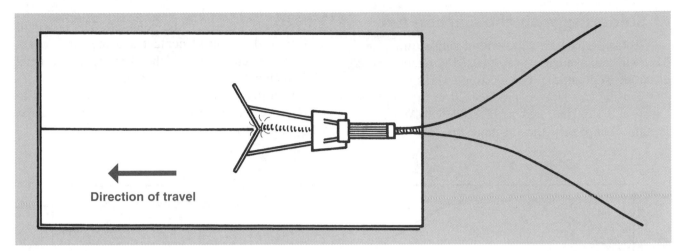

Figure 31-6.
Keep torch parallel to the joint.

6. Preheat the joint by running the flame 3 or 4″ (75 mm or 100 mm) up and down the weld joint. This will improve the weld by helping the bead to flow more smoothly.
7. Begin the weld by moving the torch to the start of the joint.
8. Place the tip of the filler rod in the arc. If the coating melts off and flows into the joint, the metal is hot enough and ready for welding. Let the filler metal melt and fuse into the joint.
9. Move the arc torch slowly along the joint, adding filler metal to the molten puddle as needed.

Single Carbon Arc Welding Operations

Single carbon arc welding operations require a DC welding machine. The machine must be set for straight polarity (DCEN). Point the electrode as shown in **Figure 31-7.** Do not attempt to use the carbon in a regular electrode holder.

Approximate current values are shown in **Figure 31-8.** Amperage is too high if the carbon burns cherry red more than 1 1/4″ (32 mm) from the tip.

The arc is struck by bringing the carbon electrode in contact with the work and withdrawing it to the proper arc length (usually 3 or 4 times the thickness of the metal).

Should the arc be broken, do not restart it in the hot weld area. Strike it a short distance beyond the weld and bring it back into the weld area.

Hold the filler rod as shown in **Figure 31-9.** Add the filler rod by feeding it into the molten pool as needed. Keep moving the weld pool along the weld joint.

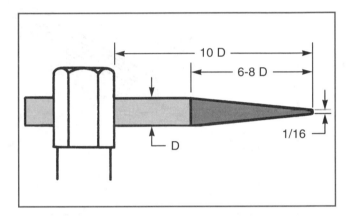

Figure 31-7.
Recommended point for single carbon arc electrode.

Carbon Diameter	Amperage Range Carbon	Amperage Range Graphite
1/8	15–30	15–35
3/16	25–55	25–60
1/4	50–85	50–90
5/16	75–115	80–125
3/8	100–150	110–165

Figure 31-8.
Approximate current values for carbon and graphite electrodes for DC machines.

Soldering with the Carbon Arc

Soldering can be done with a single carbon arc. The twin carbon arc torch can be used by removing one electrode. A pointed carbon electrode works best.

Set the machine to produce 20 to 30 amperes. Clean and flux the area to be soldered. Attach the ground clamp, turn the machine *on,* and touch the electrode on the spot to be soldered. The area will heat up at the point of contact and melt the solder. Move the electrode across the work with one hand while manipulating the solder with the other.

As no arc is developed, it is not necessary to wear a face shield when soldering. However, regular safety goggles must be worn.

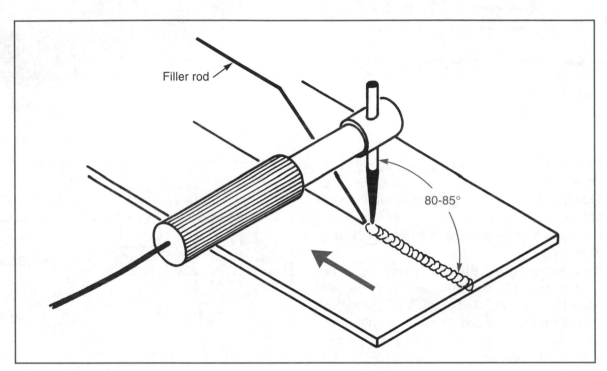

Filler rod

80-85°

Figure 31-9.
Position of torch and filler rod when welding with the single carbon arc torch.

Name _____ Score _____

Check Your Progress

1. Which of the following statements are characteristic of carbon arc welding? Check correct answer(s).

    a. _____ Produces a flame type heat.

    b. _____ No metal is transferred to the weld by the arc.

    c. _____ Metal is transferred to the weld by the arc.

    d. _____ None of the above.

2. The electrodes are made from _____ and _____. The _____ electrode is most expensive.

3. Why must a head shield be worn when welding with a twin carbon arc torch?

4. Make a sketch of a properly sharpened electrode.

5. The _____ welding machine is used for single carbon arc welding. It should be set for _____ polarity.

6. For best results, never attempt to use the electrode in a _____.

Things to Do

1. Repoint a carbon electrode.

2. Adjust the carbons on a twin carbon arc torch.

3. Use the twin carbon arc torch to make a butt weld in 1/8″ (3 mm) thick aluminum.

4. In the space below, make sketches of three joints that can be made with a carbon arc torch without filler metal.

5. Demonstrate soft soldering with the carbon arc torch.

6. Practice making welds with the twin carbon arc torch until you can produce an acceptable joint using filler rod.

7. Practice making welds with the single carbon arc torch without using filler rod until you can produce an acceptable joint.

8. Practice soft soldering with the carbon arc torch until you can produce a satisfactory soldered joint.

Unit 32
Arc Welding Aluminum

After completing this unit, you will be able to:
- ○ Explain why aluminum work material should be preheated.
- ○ Describe the importance of removing slag on an aluminum weld.
- ○ Successfully arc weld two pieces of aluminum.

An entire family of metals is identified by the term *aluminum.* They range from pure aluminum (99.45% pure), which is extremely soft, to alloys that are pound-for-pound stronger than structural steel.

Diameter	Amperage
3/32″	40–80
1/8″	80–130
5/32″	100–150
3/16″	125–200
1/4″	160–240

Figure 32-1.
Sizes and recommended amperages for aluminum arc welding rod. Amperage may vary slightly with different manufacturers' rods.

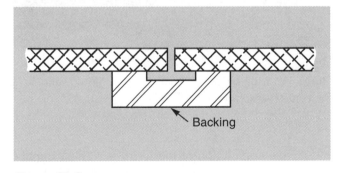

Figure 32-2.
Use backing if complete penetration is to be obtained with no burn-through.

Aluminum is a nonferrous metal. A *nonferrous metal* contains no iron, except possibly in very small quantities as impurities.

Welding Procedure

It is possible to arc weld many of the aluminum alloys used to make sheet, plates, and castings. The type of electrode most frequently employed is a heavy flux-coated rod composed of 95 percent aluminum and 5 percent silicon.

The flux coating is designed to dissolve the oxide film on aluminum and to give stability to the arc. The coating also prevents excessive oxidation during cooling.

Electrodes for arc welding aluminum give best results when used with DC reverse polarity. Recommended amperage settings are shown in **Figure 32-1.**

Aluminum sheet less than 1/8″ (3 mm) thick is difficult to arc weld because of problems in controlling the arc at low current settings. Thin sections of aluminum are usually joined using the GTAW or GMAW processes.

Little preparation is generally needed when welding on aluminum because it is easy to get good penetration. Backing is recommended if complete penetration is to be obtained without burning holes, **Figure 32-2.**

Typical joint designs are shown in **Figure 32-3.**

Because aluminum has such high heat-conducting properties, work should be preheated to 250°F to 400°F (121°C to 204°C) to maintain the weld pool. Preheating also avoids rapid cooling which can cause porosity in the weld. It also helps minimize distortion.

Strike the arc using the "scratch method." This is the only means of striking an arc on aluminum without sticking. Hold a short arc, about 1/8″ to 3/16″ (3 mm to 5 mm) long. A long arc is difficult to control and often results in excessive spatter.

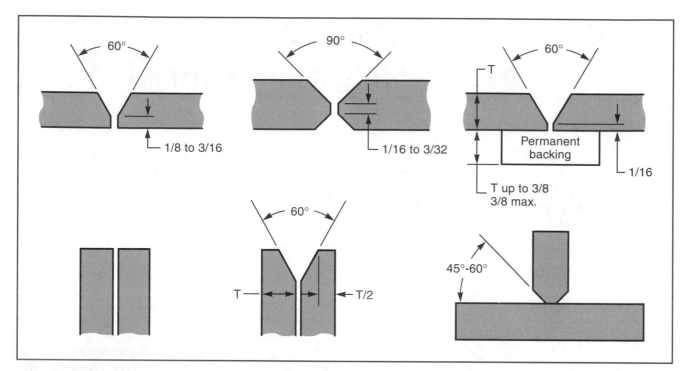

Figure 32-3.
Typical joint designs recommended when welding aluminum.

Keep the electrode in a near vertical position, **Figure 32-4.** Porosity may result if it is tipped too far forward.

Move the electrode along the joint at a uniform rate so that an even bead is deposited, **Figure 32-5.** Make the pass in a straight line. Do not use a weave pattern.

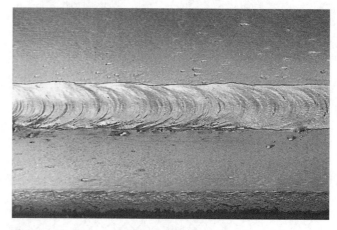

Figure 32-5.
Butt joint made on aluminum alloy plate. Note evenness of bead. (Hobart Brothers Co.)

Upon completion of the weld, it is most important to remove the slag. Slag left on the weld will attack the aluminum and damage its surface. Use the cleaning method recommended by the manufacturer of the rod you are using. Regardless of how the weld is cleaned, rinse the weld area thoroughly with warm water.

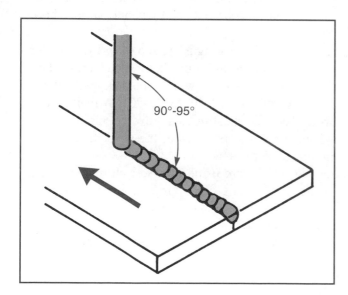

Figure 32-4.
Keep electrode in a near vertical position.

Name _____   Score _____

Check Your Progress

1. Aluminum is a _____ metal because it does not contain _____.

2. Many people think of aluminum as a metal that is light and strong. Actually, it is a _____ of metals ranging from _____ to aluminum that is _____.

3. The rod most frequently used to arc weld aluminum is composed of _____ percent _____ and _____ percent _____.

4. Why is backing recommended when complete penetration of the joint is required?

   _____

   _____

   _____

   _____

5. It is recommended that aluminum be preheated before welding because heating: Check correct answer(s).
   a. _____ helps minimize distortion.
   b. _____ helps the weld cool faster.
   c. _____ helps avoid rapid cooling which causes porosity.
   d. _____ All of the above.
   e. _____ None of the above.

6. Strike the arc using the _____ method. Otherwise, the electrode will _____.

7. Why should the slag be removed from the weld?

Things to Do

1. Secure examples of welds made in aluminum.

2. Several other forms of welding are used to weld aluminum. Research these different types and prepare a report on them. Hint—What are TIG and MIG welding operations?

3. Prepare a list of electrodes recommended by the different manufacturers for arc welding aluminum. Secure samples of these electrodes.

4. Practice arc welding aluminum until you can produce an acceptable joint. Remove a section from the practice piece and break or bend it until it is flattened and examine for weld integrity.

5. Make backing strips or bars suitable to be used when welding aluminum.

Notes

Unit 33
Welding Pipe

After completing this unit, you will be able to:
- ○ Identify the three positions used in pipe welding.
- ○ Use the roll-welding technique to run a smooth and uniform bead around a pipe.
- ○ Make open and closed butt joints that have even penetration without burn-through.

Great quantities of pipe are used for industrial, commercial, and residential applications. Arc welding has become an accepted technique for joining lengths of pipe together because it is the easiest and simplest method, especially in the larger pipe sizes. Almost all oil and gas pipelines are arc welded.

This unit will serve as an introduction to pipe welding. Pipe welding utilizes the three fundamental pipe positions shown in **Figure 33-1:** horizontal fixed, horizontal rolled, and vertical.

The horizontal rolled position is preferred whenever possible, because all welding is done in the flat or downhand position. The horizontal fixed position is the least desirable because of the overhead welding required. This is the most difficult position in which to weld.

Running Practice Beads on Pipe by Roll Welding

There are several kinds of pipe—cast iron, alloy steel, mild steel, etc. The pipe used for the exercises in this unit is 4–6″ (101–152 mm) diameter standard weight mild steel pipe. Pipe size indicates the internal diameter in standard weight pipe from 1–12″ (25–305 mm) diameter.

Make the welds with 1/8″ (3 mm) diameter E6010, E6011, or E7018 electrodes. This meets the standards established by the welding industry for welding pipe, which will be subjected to pressure.

Running Beads on the Circumference of Pipe

This exercise is designed to give you practice in running beads on the circumference of pipe. The roll welding technique is employed.

Secure a piece of 4–6″ (101–152 mm) diameter pipe 12–15″ (305–381 mm) long, **Figure 33-2.**

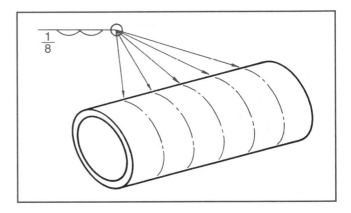

Figure 33-2.
Pipe welding problem. Run stringer beads on a section of pipe, using 4–6″ diameter pipe 12–15″ long.

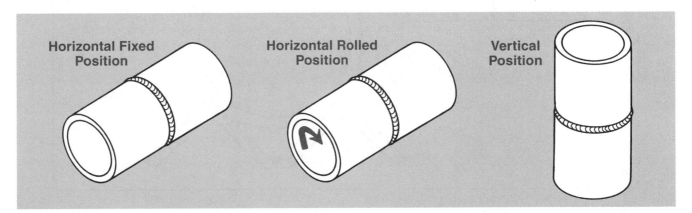

Figure 33-1.
The three basic positions used to weld pipe.

To be sure that the bead at the finish will be aligned with the start of the bead, draw guidelines on the circumference of the pipe with soapstone or chalk. This can be done by using the edge of a steel tape measure wrapped around the pipe and used as a guide.

Support the pipe so it will not roll on the welding table. The advantage of roll welding is that the welding is done in the flat or downhand position. Start the weld at about 10 o'clock position and run the bead to about the 2 o'clock position, **Figure 33-3.**

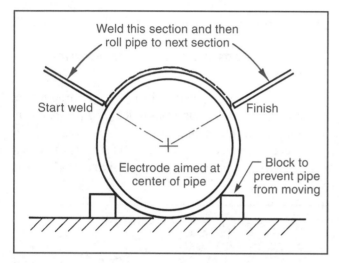

Figure 33-3.
In roll welding pipe, the weld is started at approximately 10 o'clock and is run to approximately 2 o'clock before the pipe is rolled to permit the next section to be welded.

Control electrode travel so that a smooth, uniform bead is formed. When you complete the first section, roll the pipe and run the next section. Repeat until the bead has been deposited around the pipe.

Practice running beads around the pipe until a satisfactory weld is produced. There should be little indication in the bead where the weld was stopped and restarted after the pipe was rolled to a new position.

Roll Welding a Closed Square Butt Pipe Joint

Secure two pieces of 4–6″ (101–152 mm) diameter pipe, each about 3″ (76 mm) long. Use 1/8″ (3 mm) diameter E6010, E6011, or E7018 electrodes.

Align the two sections of pipe in a length of angle iron, then tack weld at three or four equally spaced points around the joint, **Figure 33-4.**

Join the sections by means of the same technique used to run practice beads in the previous exercise. However, use extreme care to get uniform penetration without burning through or having "icicles" form on the inside of the joint.

When you can produce a satisfactory closed butt joint, practice making open butt joint welds on the same size pipe. Position the sections with a space between them equal to about one-half the diameter of the welding rod being used. Use care to get uniform penetration without burn-through.

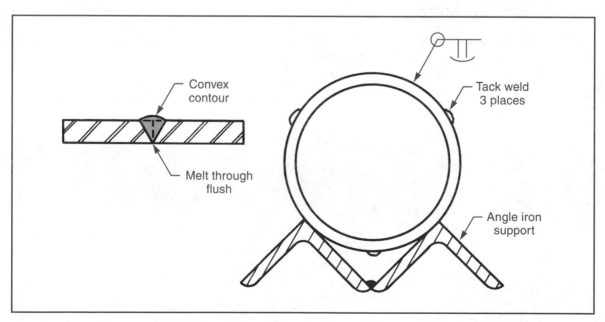

Figure 33-4.
Use sections of angle iron to hold the pipe in alignment until it is tack welded in at least three places. The required weld is shown at the left.

After you can produce satisfactory closed and open butt joints on pipe, prepare additional pieces of pipe for roll welding beveled butt joints, **Figure 33-5.**

Align the sections and tack weld in place. Make the root bead with 1/8″ (3 mm) diameter E6010 or E6011 electrodes. Use care to get proper penetration. Run the remaining beads with 1/8″ or 5/32″ (3 mm or 4 mm) diameter E6010, E6011, or E7018 electrodes.

Make additional practice pieces until a satisfactory weld can be produced.

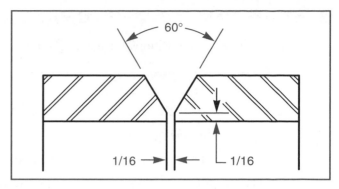

Figure 33-5.
Joint preparation for roll welding a beveled butt joint on pipe sections.

Check Your Progress

1. List the three basic positions for welding pipe.

   a. _____

   b. _____

   c. _____

2. The _____ position is preferred because _____.

3. The welding rod _____ will meet the standards of the welding industry for welding pipe.

4. When the preferred welding position is employed, the weld should start at about _____ o'clock and the bead is run to about the _____ o'clock position.

5. When welding pipe, extreme care must be taken to get _____ and to prevent _____.

Things to Do

1. Secure samples of the various types of pipe that are frequently joined by arc welding.

2. Secure a section of pipe with a welded joint made by a professional welder of pipe.

3. The term "stovepipe" is used to indicate a pipe welding position. Research the term and describe what it means.

4. Research the requirements for becoming certified as a pipe welder.

5. Make a careful visual examination of the pipe joints you have welded. Then cut a section 1–1 1/2″ (25–38 mm) wide from the weld and perform the bend test described in previous units.

6. Secure two sections of pipe and, after aligning and tack welding them, make the weld while the sections are in a vertical position.

7. Prepare additional pipe sections and practice welding the joint while they are in the horizontal fixed position.

8. Several types of clamps, fixtures, and rings are used for positioning and holding pipe while it is welded. Secure information on them from commercial sources.

9. Several types of pipe joints are used—*tee, offset, butt, branch, two joint elbow,* and *sealed end.* Research these joints and prepare sketches of them.

Unit 34
Identifying Metals

After completing this unit, you will be able to:
○ List the four general metal classifications.
○ Explain how the spark test is used to determine the grade of steel.
○ Describe how a mill file is used to determine the degree of hardness of steel.

Industry uses more than a thousand different metals. A large number of them can be welded by one or more of the welding processes.

With so many metals available, the welder often has great difficulty identifying the metal so that the proper electrode can be selected. In general, the metals used today can be classified as:

- *Ferrous metals.* Metals that contain iron in their composition. These metals are cast iron, steel, etc.
- *Nonferrous metals.* These metals contain no iron and include copper, lead, tin, aluminum, etc.
- *High temperature metals.* Metals in this category have the unique properties of high strength for extended periods at elevated temperatures. This group includes nickel base alloys, molybdenum, tantalum, etc.
- *Rare metals.* These metals are not available in large quantities and are usually very costly, often several times that of gold. Included in this group are such metals as cerium, europium, holmium, etc.

However, the average welder cannot be expected to be familiar with the various characteristics and welding techniques best suited to join each of these metals.

Because the majority of the welding problems in this workbook are concerned with ferrous metals, only this category of metals will be covered in this unit.

Since many metals are similar in appearance, how can the welder tell, with the limited facilities available, whether the metal is a ferrous metal? And, after establishing that it is a ferrous metal, how can it be determined whether the metal is a mild steel, alloy steel, or the like?

Identifying Iron and Steel

The welder has several tests available. With experience, you can identify metals to a remarkably accurate degree.

If prints or drawings are available, just refer to the title block for the type of metal to be used, **Figure 34-1.**

A magnet can be used to separate iron and steel from nonferrous metals. Almost all steels are affected by magnetism, while nonferrous metals are not (some stainless steels are not magnetic).

The *spark test,* **Figure 34-2,** is also used to determine the grade and carbon content of steel. It has proven successful as a practical shop test. To test steel by this method, lightly touch the piece to a grinding wheel and observe the resulting sparks. For best results, the area should be darkened.

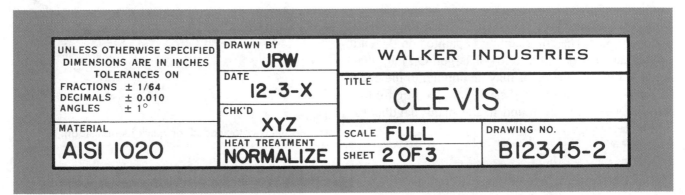

UNLESS OTHERWISE SPECIFIED DIMENSIONS ARE IN INCHES TOLERANCES ON FRACTIONS ± 1/64 DECIMALS ± 0.010 ANGLES ± 1°	DRAWN BY JRW	WALKER INDUSTRIES	
	DATE 12-3-X	TITLE CLEVIS	
	CHK'D XYZ		
MATERIAL AISI 1020	HEAT TREATMENT NORMALIZE	SCALE FULL	DRAWING NO.
		SHEET 2 OF 3	B12345-2

Figure 34-1.
Title block found on most drawings.

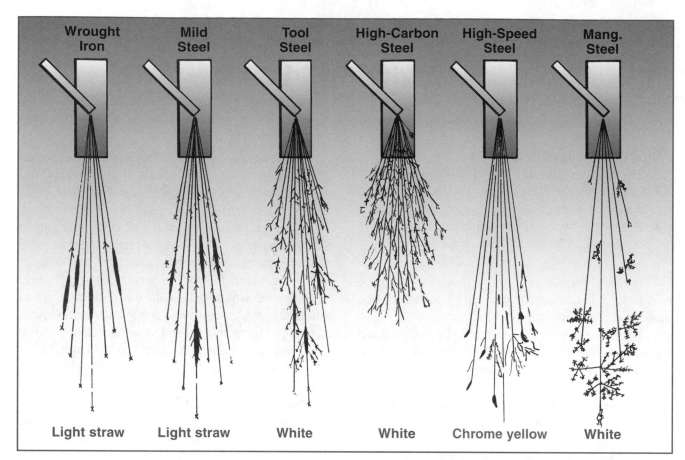

Wrought Iron	Mild Steel	Tool Steel	High-Carbon Steel	High-Speed Steel	Mang. Steel
Light straw	Light straw	White	White	Chrome yellow	White

Figure 34-2.
Spark test used to determine grades of steel.

The welder can use a mill file to ascertain the degree of hardness of steel. The technique works as follows:

- Mild steel can be removed easily.
- Medium carbon steel can be cut with moderate pressure.
- High alloy steel can be cut, but with difficulty.
- Tool steel requires great pressure to cut.
- Hardened tool steel cannot be cut.

If the welder draws metal from the storage area, he may use *color coding,* **Figure 34-3,** which is another method of identifying steel. Each different kind of commonly used steel is identified by a specific color. The color coding is painted on the end of the bar, if it is 1″ (25 mm) or larger. On bars that are less than 1″, the color may be applied to the ends of the bar or on an attached card. Color coding for the various steels can be found in the many handbooks available.

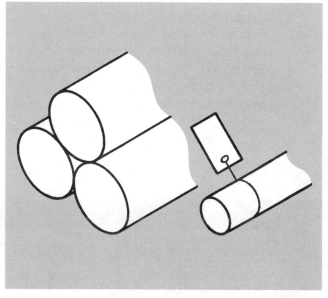

Figure 34-3.
How steel is color coded for quick identification.

Name _____ Score _____

Check Your Progress

1. Why must the welder know what metal he is welding?

   _____

2. List the four general classifications of metals:

   a. _____

   b. _____

   c. _____

   d. _____

3. Industry uses more than a _____ different metals.

4. _____ contain no iron.

5. _____ contain iron.

6. List three methods that the welder can use to help identify the metal he is to weld. Describe each briefly.

   a. _____

   b. _____

   c. _____

Things to Do

1. There are other techniques employed to determine types of metal but they require equipment that the welder does not normally have access to on the job. Research these methods and prepare a paper that will describe two of them.

2. Every day, you come in contact with many different metals. List them in the appropriate column below.

Ferrous Metals	**Nonferrous Metals**
_____	_____
_____	_____
_____	_____
_____	_____
_____	_____
_____	_____
_____	_____

Notes

Unit 35
Quality Control

After completing this unit, you will be able to:

○ Explain the differences between destructive and nondestructive testing.

○ Summarize each of the four nondestructive testing methods.

○ Describe guided bend testing and tensile testing.

The welds used in some applications are so critical that weld failure could be disastrous. One such application is the nuclear submarine, **Figure 35-1.** The hull must withstand tremendous water pressure at great ocean depths. It was fabricated by welding.

Tests had to be devised and developed to check the welds so that they would meet exacting design requirements. To ensure this reliability, a considerable portion of industry's budget is spent in the area of *quality control.*

Figure 35-1.
The hull of the nuclear submarine U.S.S. George Washington was fabricated by welding. Each weld was carefully checked to make sure it met or exceeded specifications. (General Dynamics Corp., Electric Boat Div.)

Basic Classifications of Quality Control Techniques

Quality control techniques fall into two basic classifications:

- *Nondestructive testing.* Testing is done in such a manner that the usefulness of the part is not impaired.
- *Destructive testing.* The part is destroyed during the testing program.

Nondestructive Testing

Nondestructive testing is a basic tool of industry. It is used where the performance of *each* weld is *critical.* By using one of the many nondestructive tests, it is possible to check each weld individually.

Destructive Testing

Destructive testing is a costly and time-consuming quality control technique. Common testing methods include tensile testing, impact testing, and torsion testing. The guided bend test is used to test welds made by welders who need to be certified (qualified). Service testing is also a destructive test. This type of test subjects the specimen to service conditions until failure or until the test time has expired. Valuable information about the material or part is provided through destructive testing.

Methods of Nondestructive Testing

Many familiar testing techniques in this category—weighing, measuring, and visual inspection—are very limited in their effectiveness. They cannot be used for checking weld reliability, weld strength, or improving weld quality.

Radiographic (X-Ray) Inspection

Inspection by radiographic means involves the use of X and gamma radiation projected through an object under inspection onto a film, **Figure 35-2.** The developed film shows an image of the internal structure of the part.

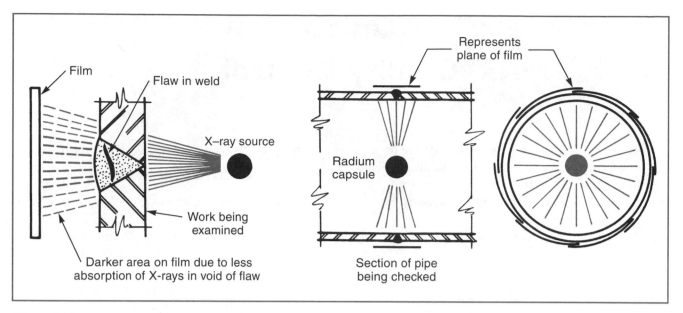

Figure 35-2.
How radiographic (X-ray) inspection works to find flaws in a weld.

Magnetic Particle Inspection

Magnetic particle inspection, **Figure 35-3,** is a method of nondestructive testing to detect flaws on or near the surface of ferromagnetic (highly magnetic) materials. The process is commonly known as *magnaflux.*

The technique involves a two step operation:
1. A magnetic field is set up electrically within the part.
2. Fine iron particles are blown (dry method) or flowed in liquid suspension (wet method) on the part.

The presence of the defect causes a break in the lines of the magnetic force field; the powder causes a visible indication of the location and size of the flaw, **Figure 35-4.**

The parts must be demagnetized (have the magnetism removed) after inspections.

Fluorescent Penetrant Inspection

The theory of fluorescent penetrant inspection is based on capillary action. The penetrant solution is applied to the surface of the part by dipping, spraying, or brushing. Capillary action literally pulls the solution into defect. The surface is rinsed clean, and before drying, a wet or dry developer is applied. This acts as a blotter and draws the penetrant back to the

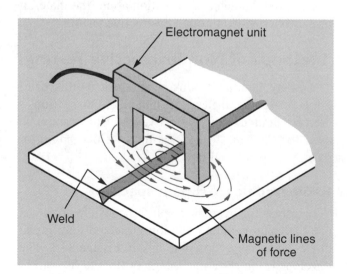

Figure 35-3.
Magnaflux (A trademark of Magnaflux Corp.) is a method of nondestructive testing to detect flaws on or near the surface of welds.

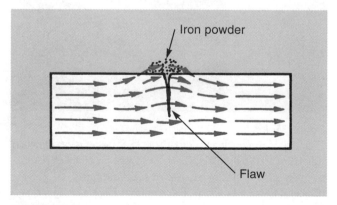

Figure 35-4.
The crack in a weld makes a magnetic field outside the part to hold iron powder and build up an indication of the flaw.

surface. When inspecting the surface under an ultraviolet light, every detail glows with fluorescent brilliance. A glowing line or spot marks each defect, right on the part.

A similar process uses nonfluorescent penetrant. The penetrant is usually bright red. When the developer is placed on the surface of the part, any defect shows bright red. The nonfluorescent penetrant does not require an ultraviolet light for inspection, making the process a little more portable and easier to use in tight places.

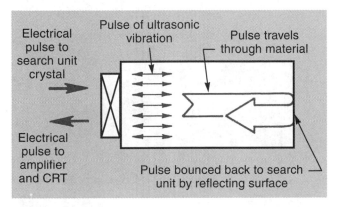

Figure 35-5.
How sound waves are used to locate flaws in welds.

Ultrasonic Inspection

Ultrasonic test equipment may be used for the nondestructive detection of flaws in almost any kind of material that is capable of conducting sound.

The human ear can hear sound waves whose frequencies range from about 20 to 20,000 cycles per second. Sound waves with a frequency greater than 20,000 cycles per second are inaudible and are known as ultrasound. Ultrasonic testing equipment uses waves of millions of cycles per second. The term "megahertz" (Mhz, meaning millions of cycles per second) is used, **Figure 35-5.**

Sound waves are used to obtain information about the interior of material by observing the echoes, which are reflected from inside the material, **Figure 35-6.** It is possible to judge the distances by the length of time required to receive an echo from a flaw. The echoes are shown on a cathode ray tube (CRT), which is very similar to a TV picture tube.

Methods of Destructive Testing

As previously mentioned, valuable information about a material or part is provided through destructive testing. The following sections detail two of the most common destructive testing techniques.

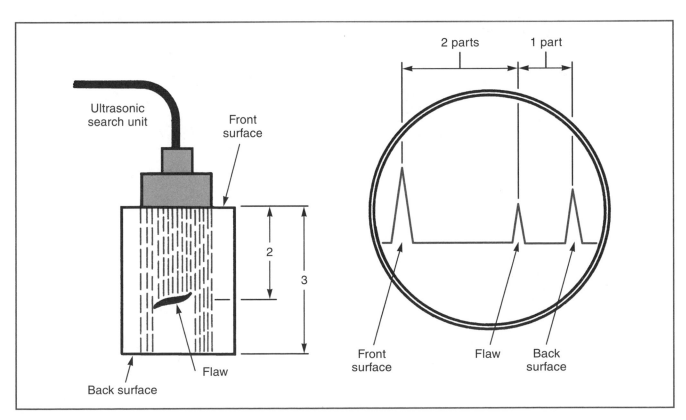

Figure 35-6.
Locating and detecting a flaw in a weld.

Guided Bend Testing

Guided bend testing is a common destructive test for both weld and welder testing. The guided bend test shown in **Figure 35-7** is used to qualify welding procedures that are required by welding codes. A welding procedure is a written description of how a weld must be made on certain products. The test is also used to qualify welders that weld products required to conform to a welding code.

Tensile Testing

Tensile testing is used to determine the strength of a material, such as the filler metal used to make a weld or the base metal used for the weld joint. The test pulls a specimen along a single axis until it breaks. **Figure 35-8** shows a tensile specimen in a testing machine. Electrode is classified by tensile strength. For example, an E7018 electrode needs to have a minimum tensile strength of 70,000 pounds per square inch (psi).

Figure 35-7.
After undergoing a guided bend test, the specimen is examined for cracks and other defects.

Figure 35-8.
During a tensile test, forces are applied to the specimen to break it into two pieces. The force applied to break the specimen is noted.

Name _____ Score _____

Check Your Progress

1. Why are quality control techniques necessary?

   _____

   _____

2. Define *destructive testing.* _____

   _____

   _____

3. Define *nondestructive testing.* _____

   _____

   _____

4. Why is destructive testing seldom used?

   _____

   _____

5. Why is nondestructive testing frequently employed?

   _____

   _____

6. List four nondestructive quality control techniques and briefly describe each.

   a. _____

   _____

   b. _____

   _____

   c. _____

   _____

   d. _____

   _____

7. Briefly describe tensile testing. _____

   _____

   _____

   _____

   _____

Things to Do

1. To assure quality welds, tests have been devised to determine the abilities of welders. Research them and present your findings to the class. (These tests can be found in many welding textbooks and reference books.)

2. Visit a manufacturing plant that makes use of welding in producing their products. Describe the quality control techniques they employ to check the reliability of welds used on their products.

3. Fabricate a device using a hydraulic jack to make guided bend tests of welds. Plans for this device can be found in the *Procedure Handbook of Arc Welding Design and Practice* by The Lincoln Electric Co.

Notes

Unit 36
Careers in Welding

After completing this unit, you will be able to:
- ○ Identify the skills that separate the various welding occupations.
- ○ Compare occupation descriptions with the corresponding educational requirements.
- ○ List resources available for gathering information about welding jobs.

Welding, in a variety of forms, is a very important tool of modern industry. So many of the things we use today employ welding in their manufacture. For example, there are more than 8,500 spot welds in the modern automobile plus many welds made by other processes.

New welding processes, electrodes, and power supplies are developed each year, and highly skilled individuals are needed to apply each.

With so many opportunities becoming available, you may want to take advantage of the job opportunities in the welding industry.

There is a broad range of requirements for jobs in the welding industry. Some require no more than a high school education while a college education is necessary to qualify for others.

The *American Welding Society* lists the following welding occupations and the minimum requirements needed to qualify for them.

Welding operator—Usually operates automatic welding equipment, which is set to produce the desired weld. The job involves monitoring the machine and reporting any change in its operation.

The operator can be trained in a short time and the job requires only a high school education. It is the least skilled of the welding jobs.

Welder (manual)—One to two years training in a vocational school is recommended. It is the job for those who like to build things with their hands and who like variety in their work. Many different industries require these skills.

Opportunities are such that a good welder can pick a particular job, the industry desired to work in, and, in many instances, the climate in which to live and work.

Welder-fitter—In addition to the skills of the manual welder, the welder-fitter must be able to plan and set up work and be able to work with welding fixtures.

A working knowledge of mathematics and the ability to read prints are essential for this level of work. A minimum of high school graduation is recommended.

Specialist welder—After becoming proficient as a welder, the craftworker may specialize in an area that is a challenge. This preference can be a welding process or a difficult-to-weld metal. This calls for study and experimentation until the technique is mastered. A vocational school or community college education is recommended.

Welding supervisor—A supervisor usually has worked as a welder-fitter and has superior skill and knowledge in that area. The welding supervisor is in charge of a team of welders. Much responsibility goes with this job.

A high school diploma or its equivalent is required for this level of work. If promotion is desired, the welding supervisor will probably have to attend additional classes.

Welding analyst—Persons in this post analyze the job and figure out requirements and estimate costs. Accuracy and exactness are required. The holder of this position must have a good working knowledge of mathematics and an extensive background in welding. Graduation from high school or its equivalent is recommended.

Welding technician—A technician stands between the welding engineer, who originates the work, and those who are concerned with completing the work. Welding technicians frequently assist in planning and development work. They use drawing instruments, collect data, perform laboratory tests, build, supervise, and control machinery and testing equipment. Many of their functions would otherwise require the services of an engineer.

The position of technician requires high school graduation or its equivalent and two years of training at a technical institute or community college. Technicians must be prepared to take additional college level courses while on the job. Persons with extensive welding experience and proven ability are also considered for this level of employment. Many technicians become engineers.

Inspector—An inspector makes sure that welding quality is maintained. Work that does not meet the standards established for the job is rejected.

The inspector must be able to read prints, be familiar with all welding standards and codes, be able to do elementary drafting and sketching, have a thorough knowledge of standard welding symbols, know how to operate and interpret the various testing equipment, and be able to assume much responsibility. An inspector must also know how to work well with people.

High school or vocational school training or their equivalent is required.

Welding group leader—This position is the highest a skilled worker can progress to without further education at the college level. It is considered part of management.

The group leader sees that the work is done efficiently and economically. It is an extremely responsible position. To do the job well, the welding group leader must know welding and must know how to direct the other welders. A high school education or its equivalent is required.

Welding engineer—This is stimulating and interesting work. Engineers are constantly faced with solving problems and are concerned with improving the quality and performance of the product. In doing this, they may develop new metals and new techniques to weld them. No matter how strong a metal may be, or how good its properties, it may have little commercial use until a way is developed to weld it. The majority of engineers are college graduates. Many have a master's degree or a doctorate.

Welding engineer (chief)—The need for welding engineers increases as the use of welding expands.

A chief welding engineer must understand metallurgy, be part design engineer, have a working knowledge of electrical engineering and mechanical engineering and, of course, have a thorough knowledge of welding and how it affects metals.

The chief welding engineer is a key person in the manufacturing sequence. A college education or equivalent experience is the minimum requirement for this job.

Sales engineer—When welding equipment is sold, the sales engineer may be called on to demonstrate the product, determine the cause of defects or malfunction of equipment in use, and find solutions to welding problems for customers.

Sales engineers must be able to talk the language of the welding industry and thoroughly understand the capabilities of the machines they sell and those of their competitors. A high school education is the minimum education for this position, although a college degree is preferred.

A welder using special welding equipment fabricating a giant tanker at the Bethlehem Shipbuilding Corp.

Name _____   Score _____

Check Your Progress

1. What welding occupation is described below? _____
   Job for someone who likes to build things. Requires one to two years training in a vocational school. Many different industries need workers with these skills.

2. Which of the following skills does the welder-fitter need? Check correct answer(s).
   a. _____ Needs skills of manual welder.
   b. _____ Must be able to read prints.
   c. _____ Have a working knowledge of mathematics.
   d. _____ Be able to plan and "set up" the work.
   e. _____ None of the above.

3. Describe the responsibilities of the specialist welder.

   _____

   _____

   _____

4. What welding occupation is described below? _____
   Usually operates automatic welding equipment. Can be trained in a short time. Job requires only a high school education.

5. Which of the following skills are needed by the welding analyst? Check correct answer(s).
   a. Figures job requirements.
   b. Estimates how much job will cost.
   c. Stands between the welding engineer and those who do the job.
   d. Makes sure that welding quality is maintained.
   e. All of the above.

6. What are the job requirements of the welding group leader?

   _____

   _____

Things to Do

1. Invite a welding instructor from a local technical college to speak to your class on employment opportunities in the field of welding in your community.

2. Invite an experienced welder to talk to your class about on-the-job conditions, what is expected by management, job responsibilities, and other points pertinent to the position.

3. Visit a shop that makes extensive use of welding in the manufacture of the products it makes.

4. Find more information on welding jobs by contacting union offices, using information in the files of the guidance office or the *Occupational Outlook Handbook*.

Conversion Table
US Conventional to Metric

When you know: ⬇	Multiply by: * = Exact		To find: ⬇
	Very accurate	Approximate	
Length			
inches	* 25.4		millimeters
inches	* 2.54		centimeters
feet	* 0.3048		meters
feet	* 30.48		centimeters
yards	* 0.9144	0.9	meters
miles	* 1.609344	1.6	kilometers
Weight			
grains	15.43236	15.4	grams
ounces	* 28.349523125	28.0	grams
ounces	* 0.028349523125	.028	kilograms
pounds	* 0.45359237	0.45	kilograms
short ton	* 0.90718474	0.9	tonnes
Volume			
teaspoons		5.0	milliliters
tablespoons		15.0	milliliters
fluid ounces	29.57353	30.0	milliliters
cups		0.24	liters
pints	* 0.473176473	0.47	liters
quarts	* 0.946352946	0.95	liters
gallons	* 3.785411784	3.8	liters
cubic inches	* 0.016387064	0.02	liters
cubic feet	* 0.028316846592	0.03	cubic meters
cubic yards	* 0.764554857984	0.76	cubic meters
Area			
square inches	* 6.4516	6.5	square centimeters
square feet	* 0.09290304	0.09	square meters
square yards	* 0.83612736	0.8	square meters
square miles		2.6	square kilometer
acres	* 0.40468564224	0.4	hectares
Temperature			
Fahrenheit	* 5/9 (after subtracting 32)		Celsius

Conversion Table
Metric to US Conventional

When you know: ⬇	Multiply by: * = Exact		To find: ⬇
	Very accurate	Approximate	
Length			
millimeters	0.0393701	0.04	inches
centimeters	0.3937008	0.4	inches
meters	3.280840	3.3	feet
meters	1.093613	1.1	yards
kilometers	0.621371	0.6	miles
Weight			
grains	0.00228571	0.0023	ounces
grams	0.03527396	0.035	ounces
kilograms	2.204623	2.2	pounds
tonnes	1.1023113	1.1	short tons
Volume			
milliliters		0.2	teaspoons
milliliters	0.06667	0.067	tablespoons
milliliters	0.03381402	0.03	fluid ounces
liters	61.02374	61.024	cubic inches
liters	2.113376	2.1	pints
liters	1.056688	1.06	quarts
liters	0.26417205	0.26	gallons
liters	0.03531467	0.035	cubic feet
cubic meters	61023.74	61023.7	cubic inches
cubic meters	35.31467	35.0	cubic feet
cubic meters	1.3079506	1.3	cubic yards
cubic meters	264.17205	264.0	gallons
Area			
square centimeters	0.1550003	0.16	square inches
square centimeters	0.00107639	0.001	square feet
square meters	10.76391	10.8	square feet
square meters	1.195990	1.2	square yards
square kilometers		0.4	square miles
hectares	2.471054	2.5	acres
Temperature			
Celsius	*9/5 (then add 32)		Fahrenheit

Fraction, Decimal, and Metric Equivalents

INCHES		MILLI-METERS	INCHES		MILLI-METERS
FRACTIONS	DECIMALS		FRACTIONS	DECIMALS	
	.00394	.1	15/32	.46875	11.9063
	.00787	.2		.47244	12.00
	.01181	.3	31/64	.484375	12.3031
1/64	.015625	.3969	1/2	.5000	12.70
	.01575	.4		.51181	13.00
	.01969	.5	33/64	.515625	13.0969
	.02362	.6	17/32	.53125	13.4938
	.02756	.7	35/64	.546875	13.8907
1/32	.03125	.7938		.55118	14.00
	.0315	.8	9/16	.5625	14.2875
	.03543	.9	37/64	.578125	14.6844
	.03937	1.00		.59055	15.00
3/64	.046875	1.1906	19/32	.59375	15.0813
1/16	.0625	1.5875	39/64	.609375	15.4782
5/64	.078125	1.9844	5/8	.625	15.875
	.07874	2.00		.62992	16.00
3/32	.09375	2.3813	41/64	.640625	16.2719
7/64	.109375	2.7781	21/32	.65625	16.6688
	.11811	3.00		.66929	17.00
1/8	.125	3.175	43/64	.671875	17.0657
9/64	.140625	3.5719	11/16	.6875	17.4625
5/32	.15625	3.9688	45/64	.703125	17.8594
	.15748	4.00		.70866	18.00
11/64	.171875	4.3656	23/32	.71875	18.2563
3/16	.1875	4.7625	47/64	.734375	18.6532
	.19685	5.00		.74803	19.00
13/64	.203125	5.1594	3/4	.7500	19.05
7/32	.21875	5.5563	49/64	.765625	19.4469
15/64	.234375	5.9531	25/32	.78125	19.8438
	.23622	6.00		.7874	20.00
1/4	.2500	6.35	51/64	.796875	20.2407
17/64	.265625	6.7469	13/16	.8125	20.6375
	.27559	7.00		.82677	21.00
9/32	.28125	7.1438	53/64	.828125`	21.0344
19/64	.296875	7.5406	27/32	.84375	21.4313
5/16	.3125	7.9375	55/64	.859375	21.8282
	.31496	8.00		.86614	22.00
21/64	.328125	8.3344	7/8	.875	22.225
11/32	.34375	8.7313	57/64	.890625	22.6219
	.35433	9.00		.90551	23.00
23/64	.359375	9.1281	29/32	.90625	23.0188
3/8	.375	9.525	59/64	.921875	23.4157
25/64	.390625	9.9219	15/16	.9375	23.8125
	.3937	10.00		.94488	24.00
13/32	.40625	10.3188	61/64	.953125	24.2094
27/64	.421875	10.7156	31/32	.96875	24.6063
	.43307	11.00		.98425	25.00
7/16	.4375	11.1125	63/64	.984375	25.0032
29/64	.453125	11.5094	1	1.0000	25.4000

Service Welding Specifications

JOINT			PASS	ELECTRODE		CURRENT (±20 AMPS)				
DESIGN	AS WELDED	SIZE		TYPE	SIZE	FLAT	HORIZ	VERT. DN.	VERT. UP	OVER HEAD
		3/16	I	E6013	1/8	130	130			
			I	E6011	1/8			130		120
		1/4-1/2	1,2	E6013	3/16	220	200			
			ALL	E6024	3/16	260				
			1,2	E6011	1/8				130	130
		1/2-UP	ALL	E6013	3/16	220	200			
				E6013	1/4	330	300			
			I	E6024	3/16	240				
			2-UP	E6024	1/4	350				
			ALL	E6011	5/32				145	145
		1/4	1,2	E6013	1/8	130	130			
				E6011	1/8			130		120
		5/16-3/8	1,2	E6013	3/16	220	200			
				E6011	1/8				120	120
		7/16-UP	1,2	E6013	1/8	140	130			
			3-UP	E6013	3/16	220	200			
			3-UP	E6024	3/16	240				
			1,2	E6011	1/8			130		120
			3-UP	E6011	5/32				145	145
		1/8-1/4	I	E6013	1/8	130	130			
			I	E6011	1/8			130		120
		5/16-1/2	1,2,3	E6013	1/8	130	130			
			I	E6024	3/16	240				
			1,2,3	E6011	1/8				130	120
		9/16-UP	I	E6013	5/32	150	150			
			2-UP	E6013	3/16	220	200			
			ALL	E6024	3/16	240				
				E6011	5/32				145	145
		1/4-UP	1,2	E6013	1/8	140	130			
			3-UP	E6013	5/32	150	150			
			3-UP	E6024	3/16	240				
			1,2	E6011	1/8			130		120
			3-UP	E6011	5/32				145	145

Dictionary of Terms

A

AC (alternating current): A current that reverses direction regularly as it rises and falls.

accurate: Made within tolerances allowed.

align: Adjusting to given points.

alloy: A mixture of two or more metals fused or melted together to form a new metal.

ampere: An electrical unit that indicates the rate of flow of electricity through a circuit.

annealing: Process of heating metal to a given temperature (exact temperature and period of time the temperature is held depends on the type of metal being annealed) and cooling it slowly to remove stresses and induce softness.

arc blow: Magnetic disturbance of an arc that causes it to waver from its intended path.

arc length: Distance from the end of the electrode to the surface of the molten pool.

arc voltage: Voltage across the welding arc.

arc welding: Process of joining metals by using the heat of an electric arc, but without pressure.

as-welded: Condition of the weld metal, welded joints, and weldments after welding but before any subsequent thermal or mechanical treatment.

B

back-step welding: A welding technique in which the increments of welding are deposited opposite the direction of progression.

backing: Material (metal, asbestos, carbon, etc.) backing up a joint during welding to facilitate obtaining a sound weld root. May be strips, rings, bars, etc.

base metal: The metal to be welded.

bevel: Angle formed by a line or surface that is not at right angles to another line or surface.

Brazing: Joining metals by fusion of non-ferrous alloys that have melting temperatures above 840°F (450°C) but lower than metal being joined.

brittleness: In some respects, the opposite of toughness. The characteristic that causes metal to break easily.

burr: Sharp edge remaining on metal after cutting, stamping, or machining. A burr can be dangerous if it is not removed.

butt weld: A weld made in the joint between two pieces of metal approximately in the same plane.

C

carbon arc welding: A welding process using a carbon rod to produce the arc and heat. Filler metal may or may not be used.

carbon steel: See *low carbon steel.*

carburizing: A process that introduces carbon to the surface of steel by heating the metal below its melting temperature in contact with carbonaceous solids, liquids, or gases. After being held at that temperature for a predetermined time, the metal is quenched.

casehardening: A process of surface hardening iron-base alloys so the surface layer, or case, is made substantially harder than interior, or core.

clearance: Distance by which one object clears another object.

complete fusion: Condition in which the base metal has melted and fused with the filler material over the entire base metal surface that is exposed for welding.

concave surface: A curved depression in the surface of an object.

concentric: Having a common center.

continuous weld: A weld that extends without interruption for its entire length.

contour: Outline of an object.

convex surface: A rounded surface on an object.

counterclockwise: From right to left in a circular motion.

cover glass: Clear glass fitted over the filter lens in goggles, a hand shield, a head shield, etc., to protect the lens from spatter.

crater: Depression at the end of a weld.

D

DC (direct current): Flow of electric current in one direction only.

deposited metal: Metal that has been deposited during welding.

depth of fusion: Distance that a weld extends into the base metal from its original surface.

downhand welding: See *flat position.*

ductility: Property of a metal that permits permanent deformation by hammering, rolling, and drawing without breaking or fracturing.

E

eccentric: Not on a common center.

edge preparation: Contour prepared on the edge of a member for welding.

electrode: A bare or flux-coated wire or rod that is melted into the base metal by an electric current passing through it.

electrode holder: Device used to hold and position the electrode.

F

face of weld: Exposed surface of a weld, made by an arc or gas welding process, on the side from which the welding was done.

fatigue: Tendency for metal to break or fracture under repeated or fluctuating stresses.

ferrous: Containing iron. Denotes family of metals in which iron is a major ingredient.

filler metal: Metal added to the weld.

fillet weld: A weld approximately triangular in shape joining two surfaces approximately at right angles to each other in a lap joint, T joint, or corner joint.

fixture: A device for holding work in position or alignment while it is being welded.

flat position: Setup where welding is performed from the upper side of the joint and the face of the weld is approximately horizontal. Sometimes called *downhand welding.*

flux: Fusible material used in brazing and welding to dissolve and facilitate the removal of oxides and other undesirable substances.

fusion zone: Area of a base metal that is melted as determined by inspecting the cross-section of a weld.

G

gage: A precision tool used by inspectors for checking metal parts to determine whether or not specified limits have been maintained during the manufacturing process.

gas pocket: A cavity in a weld caused by trapped gas.

groove weld: A weld made in the groove between two members to be joined.

H

hardening: Heating and quenching of certain iron base alloys to produce a hardness superior to that of untreated material.

heat affected zone: Portion of base metal that has not been melted but its structure properties have been altered by the heat of welding or cutting.

heat treatment: Careful application of a combination of heating and cooling cycles to a metal or an alloy in the solid state to bring about certain desirable conditions, such as hardness and toughness.

high carbon steel: Steel containing at least 0.45% carbon.

horizontal position: Setup in which the weld is made in a horizontal plane and against an approximately vertical surface.

I

incomplete fusion: A weld in which there are voids between mating parts.

inspection: Measuring and checking finished parts to determine whether they have been made to specifications.

intermittent welding: Pattern of welding where the continuity of the run is broken by unwelded spaces.

L

lap joint: A joint between two overlapping metal pieces.

lay out: To locate and scribe points for machining or forming operations.

leg of a fillet weld: Distance from root of joint to toe of fillet weld.

low carbon steel: Steel containing 0.20 percent or less carbon. Also called *carbon steel.*

M

malleability: Property of metal that determines the ease with which it can be shaped when subjected to mechanical working (forging, rolling, etc.).

manual welding: Welding entirely done by hand.

melting rate: Weight or length of electrode melted in a given period of time.

N

nonferrous: Metals containing no iron.

normalizing: A process in which ferrous alloys are heated to approximately 100°F (55°C) above the critical temperature range, and then cooled slowly in still air at room temperature to relieve stresses that have developed during welding, machining, or forming operations.

O

obtuse angle: An angle of more than 90°.

off center: Eccentric, not true.

open-circuit voltage: Voltage between the terminals of a power source when no current is flowing in a circuit.

out of true: Not on center, eccentric, out of alignment.

overhead position: Setup where welding is performed from the underside of the joint.

overlap: Protrusion of weld metal beyond the bond at the toe of the weld.

P

padding: Adding metal to a weldment's surface by depositing one or more layers of weld beads.

parent metal: See *base metal.*

pass: A single welding operation along a joint or weld deposit. A weld bead results.

peening: Mechanical working of metal by means of hammer-like blows.

penetration: Distance the fusion zone extends below the surface of a part or parts being welded.

porosity: Gas pockets or voids in the metal.

post heating: Heat applied to the work after welding or cutting.

preheating: Heat applied to the work before welding or cutting.

Q

quenching: Process of rapid cooling from an elevated temperature by contact with fluids or gases.

R

reversed polarity: Arrangement of arc welding leads in which the work is the negative pole and the electrode is the positive pole of the arc circuit.

root of weld: Points at which the bottom of a weld intersects the base metal surfaces.

root opening: Spacing or separation between metal members to be joined at the root of the joint.

S

SAE: Abbreviation for the Society of Automotive Engineers.

slag inclusion: Nonmetallic solid material trapped in the weld metal or between the weld metal and the base metal.

soldering: Method of joining metals by means of a nonferrous filler metal without fusion of the base metals. It is normally carried out at temperatures lower than 800°F (427°C).

standard: An accepted base for a uniform system of measurement and quality.

straightedge: A precision tool for checking the accuracy of flat surfaces.

straight polarity: Arrangement of arc welding leads in which the work is the positive pole and the electrode is the negative pole of the arc circuit.

strain: The measure of change in shape or size of a body, compared to its original shape or size.

stress: The intensity of internal forces at a given point in a body.

string bead: Type of weld bead made without a weaving motion.

stringer bead: The initial bead, same as *root pass.*

surfacing: See *padding.*

T

tack weld: A weld (generally short) made to hold parts in proper alignment until final welds are made. Used for assembly purposes only.

tempering: A sequence in heat treating consisting of reheating quench-hardened or normalized parts to a temperature below their transformation range and holding them at this temperature for a sufficient time to produce the desired properties.

template: A pattern or guide.

tensile strength: Maximum load a piece can support in tension without breaking or failing.

tension: Stress due to forces that tend to make a body longer.

tolerance: Permissible deviation from a basic dimension.

tool crib: Room or area in a shop where tools and supplies are stored and dispensed as needed.

toolroom: Area or department where tools, jigs, fixtures, and dies are manufactured.

true: On center.

U

underbead crack: A crack made in the heat-affected zone that does not extend to surface of the base metal.

undercut: A groove melted into base metal adjacent to the toe of the weld and left unfilled by the weld metal.

uphill welding: A pipe welding term indicating that welds are made from the bottom of the pipe to the top of the pipe. The pipe is not rotated.

V

vertical position: Setup for welding in which the axis of the weld is approximately vertical.

W

weaving: A technique of depositing metal in which the electrode is moved in an oscillating motion.

welder: Person who is capable of performing manual or semiautomatic welding operations.

welding machine: Equipment used to perform welding operations.

weldment: An assembly of component parts that are joined by welding.

weld metal: The portion of a weld that has been melted during welding.

weld pool: Portion of a weld that is molten at the place where heat is applied.

wheel dresser: A device to true the face of a grinding wheel.

whipping: An inward movement of the electrode generally employed in vertical welding to avoid undercut.

working drawing: A drawing (or drawings) that gives craftsmen information needed to make and assemble a mechanism or product.

X

X-ray: A nondestructive inspection technique for detecting internal flaws in metal parts.

Index